SUICIDES AMONG FARMERS

A STUDY OF COTTON GROWER

By the same Author

Growth and Rural Transformation : A Comparative Study of the Republic of Korea and India

---

About the Author

R.M. Mohan Rao (*b.* 1941) retired in July 2001 after serving Andhra University for about three and half decades in various capacities in the Department of Cooperation and Applied Economics. Rao secured first-class first rank in M.A. (in Rural Development) and obtained Ph.D. degree from Andhra University.

He has participated in several international Conferences/Seminars held in India and abroad. He was a member of the ICSSR delegation to North Korea, founder in-charge Director of the SAARC Centre and has been the Director of the NABARD Chair Unit since 1998. He worked as an FAO-TCDC Consultant to the Government of Zambia on Agricultural Finance in the year 2000.

He has involved himself in conducting field based research studies for various organisations like the All India Rural Credit Review Committee (1969), the NABARD (1985), Indo-Dutch Project on Alternatives in Development (1984-86), Department of Rural Development, Ministry of Agriculture (1992). He has supervised 10 Ph.D. scholars and published books on Cooperation and Agricultural Finance.

SUICIDES AMONG FARMERS
A Study of Cotton Grower

Edited by
R.M. MOHAN RAO
Prof. NABARD Chair Unit
School of Economics
Andhra University
Visakhapatnam

CONCEPT PUBLISHING COMPANY, NEW DELHI-110059

Cataloging in Publication Data--DK
Courtesy: D.K. Agencies (P) Ltd. <docinfo@dkagencies.com>

Suicides among farmers : a study of cotton grower
edited by R.M. Mohan Rao.
p. cm.
Papers presented at a national seminar held at Visakhapatnam.
Includes bibliographical references.
Includes index.
ISBN 81-8069-121-7

1. Cotton farmers--Suicidal behavior--India--Andhra Pradesh--Congresses. 2. Cotton farmers--India--Economic conditions--Congresses. I. Mohan Rao, R. M., 1941-

DDC 338.173 510 954 84 21

ISBN : 81-8069-121-7

First Published 2004

Published and Printed by
Ashok Kumar Mittal
Concept Publishing Company
A/15-16, Commercial Block, Mohan Garden,
New Delhi-110059 (India)

Phones : 25351460, 25351794
Fax : 091-11-25357103
Email : publishing@conceptpub.com

FOREWORD

This book is on emerging widespread concern with farmers' suicidal response to crop failures. When large section of farmers in the past were put to financial distress caused by either repeated crop failures or severe exploitation by the traders in the input and output markets, it resulted mainly in their agitation against their adversaries and/or armed struggles on a few occasions. Suicides seem to be a new response to crises in commercial farming, the causes of which are yet to be understood. While economic crises due to market failures are not beyond comprehension, what remains to be a puzzle still is farmers' suicidal response to them. It is important to understand this new phenomenon and put in place measures to prevent such suicides.

National Crime statistics reveal that farmers constitute 13 per cent of the 80,425 suicides committed in 1995 and the two southern states of Kerala and Karnataka were at the top of the list. Andhra Pradesh, of late, appears to be in league with them. Suicides were observed first among the small plantation farmers of Kerala and Karnataka and later among the cotton farmers of Andhra Pradesh during the late 1980s and late 1990s. They were first brought to the public awareness by the print media. The Social Scientists in turn focused their attention on the recent crises in cotton farming of Andhra Pradesh.

Suicide is a final response by an individual in the event of an unexpected catastrophic adversity. Conventionally, study of suicides fall in the realm of Psychology and Sociology. It is already well documented by the western psychologists and sociologists that suicidal tendencies are dormant in a small fraction of the populations and such a trait is inherited and runs in the families like many other mental disorders. The dormant suicidal tendencies in an individual become dominant when he faces utterly hopeless crises situations. The fact that none of the suicide victims had the family history of suicides indicate that cotton farming is fraught with lethal risks which many farmers cannot withstand. Steady expansion of cotton acreage in dry tracks of in the last three decades amidst of continuing suicides by its cultivators seems to suggest that farmers risk their lives to make a quick profit hoping that it may provide an opportunity to escape from grinding poverty.

There are hardly any studies on this emerging phenomena by economists. Prof. R.M. Mohan Rao made a pioneering effort to study the recent surge in suicides by the cotton farmers of Andhra Pradesh. As the suicides by cotton cultivators are the result of a combination of economic, social and psychological factors, Prof. Rao adopted a multidisciplunary approach to unravel the causes behind the phenomena. He carried out an in-depth field investigation of suicide victims' families, supplemented by content analysis of media reports, the results of which were discussed in a seminar attended by academicians and social activists. The outcome of this monumental endeavour is the present book containing fresh insights into the causes of cotton farmers' suicides. It consists of a new type of research that is urgently needed and should be replicated by more research institutions.

I strongly recommend this book to all social scientists and activists who want to understand the causes and processes underlying farmers' suicides and to the policy-makers, involved in the design of policies to prevent such ugly incidents in a democratic society.

R. Radhakrishna
Director
Indira Gandhi Institute of
Development Research
Mumbai

PREFACE

With commercialisation of agriculture, farming is no longer a way of life and peasants' existence is no longer ruled by custom. Currently, farmers are in the endless vortex of competition with one another. They are in perpetual danger of fancing disproportionate, atmosphere between their aspirations and achievements. Considerable sections of peasantry, particularly the small ones who grow non-food crops are in a situation of restlessness and dissatisfaction driving them with suicidogenic impulses. The suicides among cotton growers in Andhra Pradesh during 1997-98 was not an isolated instance of such impulses. According to the data on "Accidental Deaths and Suicides in India", for 1995, published by the Ministry of Home Affairs, farmers accounted for 13 per cent of the total suicide victims in the country and 2/3rds of them were below 35 years of age.

Perceptions on the causes for suicides among cotton growers in Andhra Pradesh as well as the approaches to find solutions to them have been varied. The NABARD Chair Unit, Department of Cooperation and Applied Economics, Andhra University which has conducted a 'Study on the Causes for Suicides Among Farmers in Andhra Pradesh' has organized a Seminar in collaboration with the School of Economics, Andhra University on the 4th September 1998. The Seminar was organized to examine the alternative perspectives on the causes for suicides among cotton farmers in Andhra Pradesh and to explore the ways and means to mitigate this misery future.

The present Volume presents the proceedings of the Seminar. It also consists of the Inaugural Address by Prof. C.H. Hanumantha Rao, Former Member, Planning Commission, Government of India, 11 contributed papers deliberations at the two Technical Sessions and the Plenary Session as well as concluding observations.

The Seminar was organized under two broad themes. Theme-I examined the alternative perspectives on the suicides of cotton growers. This was done against the backdrop of the prospects of cotton in the economy of Andhra Pradesh and state of cotton growers in Andhra Pradesh. Theme-II discussed the related issues that have a bearing on the suicides of cotton growers from an inter-disciplinary perspective.

The participants in the Seminar consisted of noted economists, experts from disciplines of Economics, Commerce, Sociology, Psychology, Botany, farrmers' representatives, Bankers and the NABARD officers.

Prof. R.M. Mohan Rao
Chair Professor

ACKNOWLEDGEMENTS

I am grateful to Prof. C.H. Hanumantha Rao for delivering a thought provoking Inaugural Address and for his active participation in the Seminar. We are equally grateful to other dignitaries Prof. R. Radhakrishna, Former Vice-Chancellor of Andhra University, Prof. B. Sarveswara Rao, Former Vice-Chancellor, Nagarjuna University, Prof. G. Parthasararthy, Director, Institute of Development and Planning Studies, Visakhapatnam, Dr. Yalamanchali Shivaji, Ex-M.P. and Member of Central Cotton Advisory Board, for gracing the Inaugural Session and for their active participation in the deliberations.

Our thanks are also due to the Presidents of three Sessions of the Seminar, Prof. G. Parthasarathy, Prof. M. Jagadeswara Rao, Director Agro-Economic Research Centre and Dean of Academic Affairs, Andhra University, Dr. Y. Shivaji, who presided over Session-I on Alternative Perspectives and Session-II on Related Issues and the Plenary Session respectively.

I am also thankful to the Rapporteurs of the various sessions Dr. C.S. Murthy, Prof. D.S. Prasad and Dr. R. Sudarsana Rao.

I am particularly thankful to all the paper writers, and other distinguished participants from various disciplines on the campus, representatives of farmers, Bankers and the NABARD, for their active participation in making the deliberations more meaningful.

I am thankful to NABARD, Mumbai, for sharing the results of their sponsored study with the wider public at the Seminar, the University authorities to the U.G.C. and Prof. P.V. Sarma, Former Director, School of Economics for their financial assistance.

Finally, I am grateful to Prof. R. Radhakrishna, then Vice-Chancellor, Andhra University and currently Director, Indira Gandhi Institute of Development Research, Mumbai, for his support for conducting the Seminar and kindly agreeing to write the 'Foreword'.

My collegues in the NABARD Chair Unit, Dr. Ch. Appa Rao, Dr. N. Subba Rao, Sri P. Matchi Raju, Steno and Former Student Sri M. Venkateswara Rao deserve appreciation for their support in conducting fieldwork. My friend and colleague Prof. D.S. Prasad, has been a consant source of help in editing the manuscript and I profusely thank him for all his support. I also thank Sri Ashok K. Mittal, for readily agreeing to publish the manuscript.

Prof. R.M. Mohan Rao
Chair Professor

CONTENTS

PART - II

PROCEEDINGS OF THE SEMINAR

PART - I

SELECTED PAPERS

1

Prospects of Cotton in the Economy of Andhra Pradesh

NABARD CHAIR UNIT

To gloss over the untold misery and sad stories of Prakasam District of 1987-88, and 1997-98 incidents in Warangal, Guntur and many other cotton growing districts of Andhra Pradesh as instances of a rare nature is short sightedness. There is every possibility of emergence of problems of this nature, in a more severe form with the unfolding of globalisation and liberalisation policies in the near future. Thus, there is a need to take stock of the prospects of cotton in the State's economy, as it sets the tone for the future programme. Keeping this in view, an attempt is made in this Chapter, to focus on the prospects of cotton economy in Andhra Pradesh, *viz.*, profitability of cotton, reasons for preferring cotton to other crops and adequacy of the institutional set-up to promote commercial agriculture in general and cotton farming in particular.

During the triennium ending 1994-96, India accounted for 11.5 per cent of the total cotton production in the World. It occupied third place, next to China and the U.S.A. The area under cotton in the country has almost doubled from Ha. 4.93 million in 1949-50 to Ha. 9.12 million by 1995-96. It also witnessed a near five-fold rise in production *i.e.*, 2.75 million bales (of 175 kg.) to 14.25 million bales and a three-fold increase in per hectare yield from 95 kg to 266 kg during 1949-50 and 1995-96. This increase in yield has become possible, among other things by a significant rise in the area covered under irrigation from 8.2 to 34.2 per cent between 1949-50 and 1995-96.

All-India compound growth rates of area, production and yield of principal crops with the triennium ending 1981-82 as base, shows that during 1967-68 and 1996-97 cotton stood next to wheat, rape seed and mustard with an annual growth rate of 2.80 per cent among the principle crops. If one considers the period 1980-81 to 1996-97, cotton occupied the first place with 3.75 per cent growth in yield.

In terms of production also, cotton stood first with an annual growth rate of 7.04 per cent during 1991-92 and 1996-97 (Ministry of Agriculture, 1998, 132-133). The compound growth rate of area under cotton has been increasing over time, particularly from 1980's onwards, while the rate of increase is only 0.07 during 1967-68 and 1980-81, it increased to 0.54 per cent during 1980-81 and 1996-97.

The above facts clearly affirm the growing popularity of cotton among the Indian farmers particularly since 1980's as well as their zeal to increase production and productivity.

Prospects of Cotton in Andhra Pradesh

Similar to the country's position in the World, Andhra Pradesh stood next to Maharashtra and Gujarat in terms of share in area as well as production during 1996-97. It also stood next to Punjab, Haryana and Rajasthan in terms of yield per hectare. However, the unique contribution of Andhra Pradesh lies in improving the levels of productivity. Andhra Pradesh with only 14.6 per cent of cotton area under irrigation, as against near cent per cent in Punjab, Haryana and Rajasthan achieved an average yield of 312 kgs /Ha. as against 354 kgs in Rajasthan and 441 kgs in Punjab in 1996-97 (*Ibid.*, 39-40).

During 1975-76 and 1997-98, despite fluctuations in area, production and productivity, the emerging trend is one of a rise. The area under cotton increased from Ha. 3.54 lakhs during 1975-76 to Ha. 10.07 lakhs in 1996-97, the highest during the 20 years period. Production increased by four-fold, as against a three-fold rise in area during the period *i.e.*, 5.45 lakh bales to 28.00 lakh bales by 1995-96; productivity also doubled from 262 kg/Ha. in 1975-76 to 583 kg/Ha. in 1990-91 (Andhra Pradesh Cotton Association, 1998, 68). The Cotton Advisory Board's data showed that productivity of cotton in Andhra Pradesh touched a peak of 649 kg/Ha during 1994-95. This was considered as the highest productivity among states in any year in the country so far (Lal, 1998, 107). With the exception of Maharashtra, Andhra Pradesh is the only State with the lowest coverage of cotton with irrigation. The fact that Andhra Pradesh has established a record yield of cotton in the country, being a State with low per centage of cotton area covered with irrigation and relatively low share of agriculture in bank credit clearly shows that the prospects for cotton in the State are promising. It could certainly emerge as the leader to other State's, if the needed technical know-how and institutional credit is extended.

Indices of area, production and productivity of 19 major crops with 1969-70 triennium as the base, showed, cotton stood next to black gram in terms of increase in area, first in terms of productivity and production indices during 1992-93 (Citizen's Report, 1998,109). This clearly shows, Andhra Pradesh farmers' preference for a shift towards cotton, as well as their capacities to augment production and productivity.

Profitability of Cotton

To reflect on the relative profitability of cotton *vis-a-vis* other competing crops for 1997-98, the data on cost of cultivation, yield, price per quintal, gross and net returns per hectare of irrigated/unirrigated cotton, chillies and paddy, available data in Citizen's Report and Andhra Pradesh Cotton Association are compared with our study (Tables 1.1 and 1.2).

As the data relate to an abnormal year of acute crop loss for cotton and showed a net loss for all the three crops, it is difficult to draw any meaningful inferences. The cost of cultivation per hectare of chillies is higher than that of cotton. If one considers a three year period cotton is likely to emerge as the most profitable crop compared to chillies for two reasons. Firstly, chillies are not covered under the minimum support price policy and the price fluctuation in the case of chillies is more volatile compared to cotton. Secondly, the post-harvest costs as well as losses due to de-colourisation are severe and hence it is not preferred by the farmers.

Table 1.1 : Cost of Cultivation of Cotton (per hectare) as seen from different studies vis-a-vis our Study; 1997-98

Item	*Our Study*		*Citizen's Report*		*A.P.Cotton Association*		
	Warangal	*Guntur*	*Warangal*	*Guntur*	*Guntur and Prakasam*	*Warangal and Karim Nagar*	*Kur-nool*
	Un-irri-gated	*Un-irri-gated*	*Irrigated*	*Un-irri-gated*	*Un-irri-gated*	*Un-irri-gated*	*Un-irri-gated*
1	2	3	4	5	6	7	8
Cost A_1	19706	20841	43558	28961	NA	NA	NA
Cost A_2	21988	26128	43558	28967	NA	NA	NA
Cost B	26507	27862	54949	34709	NA	NA	NA
Cost C	29555	30118	54949	34709	25200	22500	19760
Yield (in Qnt)	9.20	7.50	14.60	10.10	20.00	15.00	20.80
Unit Cost (Rs.1 kg)	32.13	40.16	37.64	34.37	12.60	15.00	9.50
Price (Net in Rs.)	1893	1917	1960	1960	1965	1965	1965
Goss Returns	17416	14378	28616	19796	39300	29475	40970
Net Returns	-12139	-15740	-26333	-14913	14100	6975	21210
Input-Output Ratio	0.59	0.48	0.52	0.57	1.56	1.31	2.07

Table 1.2 : Cost of Cultivation of Chillies &Paddy (per hectare) as seen from different Studies *vis-a-vis* our Study 1997-98

Item	*Chillies.*			*Paddy*		
	Our Study		*Citizen's Report*	*Our Study*		*Citizen's Report*
	Warangal Un-irrigated	*Guntur Un-irrigated*	*Guntur Irriga-ted*	*Warangal*	*Guntur*	*Warangal*
1	2	3	4	5	6	7
Cost A_1	30327	32555	52359	11468	10382	18706
Cost A_2	31711	36313	52359	12828	16098	18706
Cost B	39266	41950	63775	15595	16454	21503
Cost C	41615	45900	63775	18010	18618	21503
Yield (in Qnt)	11.60	16.10	19.50	30.60	40.00	39.00
Unit Cost (Rs. 1 kg)	35.88	28.51	32.71	5.89	4.65	5.51
Price (Net in Rs.)	2025	2150	2050	420	420	440
Gross Returns	23490	34615	39975	12852	16804	17160
Net Returns	-18125	-11285	-23800	-5158	-1814	-4343
Input-output Ratio	0.56	0.75	0.63	0.71	0.90	0.80

Source : Col. 2, 3, 5 and 6: our study 4 & 7 : *Ibid.*, pp. 128, 130

To overcome the data problem, data from the Farm Management Study for 1981-84 (Nagarjuna Sagar Area) are considered. This Study is more authentic on cost of cultivation and profitability of crops in Andhra Pradesh (Table 1.3).

Table 1.3 : Cost of Cultivation of Cotton, Chillies and Paddy, 1981-84 in Andhra Pradesh

Item	*Cotton*		*Chillies*	*Paddy*
	Irrigated	*Un-Irrigated*		
Cost A_1	7311	6867	6758	3895
Cost A_2	7815	73014	6762	4005
Cost B	10666	10008	8947	6218
Cost C	11066	10405	9837	6504
Yield (in Qnt)	22.38	19.34	15.37	41
Unit Cost (Rs. 1 kg.)	4.94	5.38	6.40	1.59
Price (Net in Rs.)	576.40	567.97	692.45	152.32
Gross Returns	12900	10985	10643	6245
Net Returns	1834	580	806	-259
Input-Output Ratio	1.17	1.06	1.08	0.96

Source : B. Prasada Rao & L.K. Mohana Rao, Studies in the Economics of Farm Management in the Command Area of Nagarjuna Sagar Irrigation Project (Andhra Pradesh), 1981-84.

Reasons for Adoption of Cotton by New Generation of Farmers in the Study Area

An analysis of the background data of the victims has sharply brought out that a majority of them were young, marginal and small farmers while some of them are land-less and illiterates. Given their economic status, and high-cost high-risk nature of cotton cultivation on one hand and their limited access to credit, inputs and extension services on the other, it is pertinent to question, what prompted them to take to cotton despite all the odds?

Further, despite the ups and downs of farming in general and cotton in particular, its cultivation is fast making in-roads into areas alien to its cultivation in the past. There is an increase in area under cotton within the districts and villages which were mute witnesses to suicides among fellow farmers. What factors contributed for this emerging craze towards cotton, despite its die-able nature, needs reflection.

In this context it must be noted that cotton is die-able. It is also equally viable or more viable than most of the competing crops under favourable soil and climatic conditions. The reasons are lucid.

Firstly, under un-irrigated conditions there is no other crop with the exception of chillies that ensures higher yields and incomes under normal conditions.

Cultivation of chillies exposes the farmer to price risk as it is not covered by minimum support prices. Further, the post-harvesting work of chillies is more taxing to the farmer while cotton is free from such problems. This encourages farmers to prefer cotton to chillies.

Many other technical reasons are also advocated by agricultural scientists and entomologists in support of farmers preference for cotton (Raghunadha Rao and Hari Prasada Rao 1998, 69-70). Some of the reasons as given by them are :

1. In the upland rainfed situations, cotton is a versatile crop that can be sown at any time from June to September without any significant impact on yield. In crops like sesamum and groundnut, if the sowings are delayed beyond the first fort night of June, the yields are drastically reduced; so also, pulses like greengram and blackgram cannot be sown after 15th of July.
2. In case of two crop (kharif & rabi) preparatory cultivation and sowing of Rabi crop depends on soil moisture and rainfall in September-October, which has its own risks. Therefore, farmers always prefer a single crop like cotton and redgram in these situations.
3. Once the crop germinates, cotton can withstand prolonged dry spell of 30 to 45 days; while other crops like sesamum, ground nut, green gram and black gram dry up.
4. After the sowings are over and germination is assured, the farmer's major responsibility lies on inter-cultivations and weed control. In case of cotton, this is easy compared to other crops.
5. Regarding plant protection measures, cotton and chillies

require the maximum attention. These measures are easier for cotton than those of chillies. Further, total investment per hectare is lower in case of cotton compared to chillies.

6. Fluctuations in market price structure are much less in cotton compared to competing crops like chillies and tobacco.
7. There are no post-harvest operations in case of cotton compared to chilies and tobacco.
8. Thus, when the overall situation is assessed cotton is considered to be less risky and more remunerative than both chillies and tobacco.

Agricultural Extension Services

Of the 1,200 extension officers in the State, those working in office are around 1,000 leaving very few for field level extension work. Cotton crop requires extension guidance during June and November and extension staff should be able to visit each village at least four to five times, at different stages in the growth of the crop.

In the changed context, extension machinery alone cannot tackle the problem. Imparting technical knowledge to farmers, incorporating a good mix of theory and practice could help ease the problem to some extent.

Similarly, guiding farmers to identify the genuinity of inputs, seeds, fertiliser and pesticides will go a long way to check both spuriousness and adulteration. Atleast training one or two volunteers in a village would be of great help in this regard. Further, testing of sample of inputs locally, by the volunteers goes a long way in eliminating the alleged collusion of the government machinery and the dealers in checking adulteration etc.

Use of computers to bridge the gap between percolation of knowledge 'from lab to land' could also be explored. The computerised information on the causes for various types of crop pests and diseases, as well as the steps to be taken can be made accessible to the progressive farmers of the village, village level agricultural functionaries or the village panchayats or similar organisations. Making expert system accessible to farmers should be explored to overcome the shortage and limitations of formal agricultural extension system.

Crop Insurance

Comprehensive Crop Insurance Scheme is in vogue in Andhra Pradesh since Kharif 1985. It is aimed at assisting farmers affected by natural calamities like droughts, floods, cyclones, pests and diseases. It is intended to provide support to promote productivity of crops like coarse cereals, pulses and oil seeds. During Kharif, it covers 14 crops like, Paddy, Jowar, Maize, Ragi, Bajra, Korra, Redgram, Greengram, Blackgram, Horsegram, Groundnut, Gingely, Sunflower and Castor. In Rabi it covers 10 out of these 14 crops, with the exception of Redgram, Castor, Jowar and Korra.

Mandal or group of mandals is the unit for crop insurance. An amount of Rs. 82 crores during Kharif 1996 and Rs. 48 crore in Rabi 1996-97 was paid as compensation to farmers. During Kharif 1997 around Ha. 280.26 lakh of cropped area by 10.90 lakhs farmers were covered under crop insurance. The compensation to be paid to these farmers has been under consideration.

Experimental Crop Insurance Scheme

The General Insurance Corporation of India (GIC) has introduced a comprehensive insurance scheme in 1985-86 in collaboration with the respective State Governments covering cereals, pulses and oil seeds. The major features of the Scheme are as follows :

1. The basis of the Scheme is an area yield guarantee. The State Government decides which crops to be covered by the Scheme. The pay-out to each farmer of an insured crop in a given year is the difference between the threshold yield and the measured average yield if the measured average falls short of the threshold.
2. The Scheme is linked to institutional credit and no farmer who is not a borrower may participate.
3. The rate of premium is two per cent of the sum insured

for Rice, Wheat and Millet crops and one per cent in respect of oilseeds and Pulses. These rates are standard across India. Even this low rate of premium is subsidised by 50 per cent in the case of small and marginal farmers. The subsidy is shared by Central and State Governments.

4. The Scheme is voluntary on a State basis.

Multiple risk insurance has failed world wide. Indian experience shows unsustainable financial shortfall and a failure of insurance scheme to even mitigate the overdues of the affected banks. The need for general increase in premia to overcome short falls and incentives to repay loans are stressed. It also stressed the need for reduction in the range of risks covered (Paul Mosley and R. Krishna Murthy, 1996).

In view of the complexity of crop insurance brought out in the earlier studies and its unviable nature in economies with marginal and small farms covering at least two prominent risks plaguing cotton *viz.*, i) natural factors, and ii) pests and diseases, should be considered for coverage under crop insurance scheme being contemplated by the Government of India from Kharif 1999. However, the guidelines announced by the Government of India in this regard *viz.*, limiting it to institutional borrowers and risk coverage only to the tune of Rs. 10,000 per farmer may not hold good in respect of cotton. A measure of compulsion for all growers become vital in view of high susceptibility of cotton to pests and diseases. Further, fixing ceiling limit to only Rs. 10,000 will not offer adequate security to marginal and small farmers. Thus, there is a need to reflect further in devising a viable scheme for cotton.

To sum up, the analysis in the Chapter has clearly brought out the prospects of cotton in the Andhra Pradesh economy as reflected in expansion in area, increase in production and productivity, available evidence on the cost of production and profitability of cotton. In addition to other factors *viz.*, flexability in sowing, absence of post-harvest chores as in the case of chillies and/or tobacco, assured minimum prices, relatively low risk on many fronts compared to chillies, show that cotton cultivation is a viable proposition and calls for effective extension. However, the existing facilities for agricultural extension are too inadequate to meet the challenges of an expanding area and widening complexities in plant protection measures.

Viable crop insurance, though not easy to make it operational covering multiple risks, is vital for increasing the prospects of cotton. The Central policy of ushering in crop insurance from Kharif 1999 hopefully will go a long way in giving a boost to cotton economy in Andhra Pradesh and in the Country at large.

REFERENCES

Andhra Bank, 1996-97, 1997-98, *Annual Action Plan*, Guntur District (Mimeo).

Citizens' Report, 1998, *Gathering Agrarian Crisis—Farmers' Suicides in Warangal District (AP)*, 1998, (Mimeo), p. 127.

Fertiliser Association of India, 1995-96, *Fertiliser Statistics*, New Delhi.,

Government of India, (Ministry of Agrl.), 1998, Agricultural Statistics at a Glance, New Delhi.

Joint Director of Agriculture, 1997-98, Guntur, Annual Production Plan, Khariff.

Paul Mosley and R. Krishna Murthy, 1996, "Can Crop Insurance Work? The Case of India", *National Bank News Review*, 12 (3), pp. 12-25.

Prasada Rao, B. and L. K. Mohana Rao, 1986, *Studies in the Economics of Farm Management in the Command Area of Nagarjuna Sagar Irrigation Project* (Andhra Pradesh), Directorate of Economics and Statistics, Department of Agriculture and Co-operation, Government of India, New Delhi.

Raghunadha Rao and Hariprasada Rao, 1998, "Prospects and Perspectives of Cotton in Andhra Pradesh", *Andhra Pradesh Cotton Association, Seminar on Cotton Cultivation & Problems of Marginal Farmers*, April 1998, (Mimeo), Guntur, pp. 65-76..

Ramesh Chand and Birthal Pratap, S., 1997, "Pesticides use in Indian Agriculture in Relation to Growth in Area and Production and Technological Change", *Indian Journal an Agricultural Economics*, Vol. 3, July-Sept., 488-498.

State Bank of Hyderabad, 1996-97 and 1997-98, *Annual Action Plan*, Warangal District, (Mimeo).

2

Suicides Among Cotton Growers in Andhra Pradesh

A Survey of Warangal and Guntur Districts

NABARD CHAIR UNIT

"If life were in itself a blessing to be prized (wrote Schopehauer) and decidedly to be preferred to non-existence, the exit from it would not need to be guarded by such fearful sentinels as death and its terrors. But who should continue in life as it is if death were less terrible" (quoted by Goldbratt, 1978, 314).

Schopehaur's observation reveals the nature and extent of social strain the victim experiences. One must experience a rather high degree of suffering to pass through the doors of suicide.

Suicide : Need for conceptual Clarity

Of the three types of suicides coined by Durkheim, *viz*, the egoist, altruist and anomic, anomic suicide is the type that increases in economic crisis. Durkheimian reasoning on this type of suicide is as follows: It affects Individuals as a result of conditions of existence in modem societies. Social existence is no longer ruled by custom; individuals are in endless competition with one another; they expect a great deal of life, they demand a great deal from it. They are in perpetual danger of suffering from the disproportion between their aspirations and satisfaction. This atmosphere of restlessness and dissatisfaction is favourable to the growth of suicidogenic impulse (Aron Raymond, 1976, 43).

It is against this Durkheimian framework of anomic suicide, the causes for the suicides among cotton growers in Andhra

Pradesh are perceived. This chapter essentially presents the results of the field survey undertaken by the NABARD Chair Unit, Department of Co-operation and Applied Economics, Andhra University, Visakhapatnam in Warangal and Guntur Districts. The sample consisted of 36 victims in each of the two districts and the survey was done during July, 1998.

This chapter is divided into three parts, Part one presents the results of the Unit's study of farmers' deaths in India; Part two highlights the results based on content analysis of media reports on suicides of cotton growers in Andhra Pradesh. Part three presents the evidence on suicides of cotton growers in Warangal and Guntur districts against the backdrop of the salient features of the two districts having a bearing on the issue.

PART I

SUICIDES OF FARMERS IN INDIA

A Macro Perspective

At the All-India level, despite achievements on many fronts during the last 50 years, old problems have not whithered away while new problems have cropped up. To illustrate, Vera Antesy's words of pre-Independence days that "Indian agriculture is a gamble in monsoons" and Royal Commission's (1928) observation that "Indian farmer lives in debt, dies in debt and bequeaths it to his children" still persists. With modernisation and commercialisation and greater use of purchased inputs Indian agriculture is no longer a way of life. These changes without corresponding progress in irrigation, apart from aggravating regional inequalities, also increased risk in Indian farming. Added to these, the farmer is also greatly exposed to the vicissitudes of market economy on the one hand and declining institutional support on the other, in the wake of new economic policy ushered in 1991.

The information revolution through media has evoked rising expectations among one and all for a better way of life. The youth among small and marginal farmers, dissatisfied with their levels of living and zeal to improve their lot are also tempted towards commercial crops like cotton, tobacco, pulses, onions etc., which are subjected to multiple risk on many fronts. The 1987-88 deaths of cotton growers in Guntur and Prakasam districts is a pointer to this. Thus, these groups are in the perpetual danger of suffering from disproportion between their expectations and aspirations and real life realities and satisfaction. This restlessness and dissatisfaction has sown seeds of suicidogenic impulses among farmers. This is reflected in the suicides among farmers across all States in India, though remained invisible for long and came to the fore with media publicity of cotton growers' suicides in Andhra Pradesh during 1997-98.

Trends in Suicides in India

An examination of the data on the incidence of suicides in India during 1978 and 1995 shows that from 1979, there is a continuous increase in the number of suicide cases. It has increased from 40,207 in 1978 to 80,424 in 1995. Suicides per lakh of population increased from 6.3 in 1978 to 9.74 in 1995. Suicide deaths due to economic bankruptcy accounted for 1.5 per cent of the total suicides in India during 1988 and 1992. This has increased to 2.11 per cent by 1995. This is certainly an underestimate as in nearly more than 1/3rd of the cases the causes are unidentified. During the same period around 4 per cent of deaths in Andhra Pradesh and 2.5 per cent of the deaths at the all-India level are traced to poverty. These shares have risen to 6.4 and 2.65 by 1995 in Andhra Pradesh and at all-India level respectively. However, the raising trends in suicide rates have not received the attention they deserve for any serious study by academics or enquiry by the Governmental Agencies.

Suicides among Farmers are prevalent even prior to the ushering in of market economy in 1991. The 1987-88 deaths of cotton farmers in Andhra Pradesh are pointers to this. However, they remained invisible for long and came to the fore only in 1997-98. Suicides' spree among farmers in Andhra Pradesh, Karnataka, parts of Vidharbha in Maharashtra and Punjab, has aroused public opinion due to unprecedented rise in the number of deaths as well as due to the active role of media in highlighting this human misery.

Suicides Among Farmers

The National Crime Records Bureau (1997) for the first time provided a break-up of professional status of suicide victims in India. According to this Report 10,720 farmers both male and female are among the total suicide deaths of 80,424 in 1995. Around 75 per cent of this group died in the prime age of 35 years. From this data, the rate of suicides of farmers (suicides per lakh of population) were computed. The rate of suicides among farmers in 10 out of the 19 States for which the break up by professional status of the victim is available, exceeded the all-India average of 9.74 for all categories of suicide victims. Among the 19 States considered, Karnataka accounted for about 1/5th and Kerala for 12 per cent of the total suicides among farmers during 1995. However, the rate of suicides of farmers is the highest in Kerala (129.90). It is also the State with 81.03 per cent of area under non-food crops, highest ratio of share of small farmers in area to operational holdings (7.90) and the highest rate of suicides among general population in the country. Its share in rural credit to total credit in the State is the second lowest, with an equally unenviable share of priority sector advances to total bank credit, and a high per centage of suicides due to economic bankruptcy.

In view of the features associated with farmers deaths in Kerala on one hand and Cohen's (1965, 6) observation that others experiences influence the behavioural pattern of others under similar type of strain, it is hypothesized that farmers' suicides are likely to be more:

1. in areas with high-risk non-food crops,
2. in areas with highly favourable ratio of land in relation to small holdings,
3. in areas with high rate of suicides in the general public, and
4. in areas with high percentage of deaths due to economic bankruptcy.

Since economic burden is central to small farm economy under commercial farming, variables like percentage of area under cotton, percentage of small holdings, flow of bank credit to rural areas and share of priority sector advances in bank credit are also considered.

To assess to what extent these variables have a bearing on the issue under focus, correlation and regression analysis was carried out using State level cross-section data for 19 States for which data on farmers' deaths are available. Correlation analysis showed a positive and significant relationship between the rate of suicides of farmers and 1. ratio of share in area to share in holdings of small farmers, 2. per centage of area under non-food crops, 3. rate of suicides among total population, and 4. per centage of suicides due to economic bankruptcy.

To eliminate the problem of multi-colleniarity among highly correlated variables and to identify which of these actually affect the suicide rates among farmers, a set of regression equations are estimated with various combinations of these variables. Finally, the following equations are retained.

1. $RSF = -\,22.03 + \underset{(4.28)}{8.72\ RSASHSF} + \underset{(5.70)}{2.59\ RSTP} \qquad R_2 = 0.82$

2. $RSF = 3.38 + \underset{(4.08)}{11.93\ PSTB} \qquad R_2 = 0.46$

3. $RSF = -\,14.25 + \underset{(3.47)}{1.13\ PNFC} \qquad R_2 = 0.38$

The regression results have clearly brought out that rate of suicides of farmers is positively and significantly associated with RSASHSF (Ratio of Share in Operated Area to Share in Small Farmer Holdings), RSTP (Rate of Suicides in Total Population), PSTB (Percentage of Suicides due to Economic Bankruptcy) and PNFC (Percentage of Area Under Non-Food Crops).

The positive and significant association between pesticide-use and percentage area under cotton and percentage of small holdings on the one hand and negative and significant association between percentage of area under cotton and share of rural credit in bank credit, negative and significant association between share of area under minor irrigation and share of priority sector advances to agriculture, as well as the results of the regression analysis discussed above substantiate the hypotheses on the rate of suicides among farmers in India.

The results also show that the rate of suicides of farmers are more manifest in areas with predominance of non-food crops, small holdings, minor irrigation and low share of bank credit to rural areas and low share of priority sector advances to agriculture. The problem of the unorganised and illiterate lot of small farmers who are mere price-takers in all markets is likely to worsen and may unleash suicidogenic impulses on a larger scale once market economy unfolds itself in its true colours in the near future. This calls for efforts to organise them to increase their bargaining strength on one hand and to ensure policy commitment for appropriate interventions by the Government to protect the interests of all the vulnerable sections within the framework of the market economy.

Objectives of the Study

The major objective of the Study is to identify the causes for suicides among farmers raising cotton in Andhra Pradesh. The specific objectives of the Study are:

(i) To get at an overview of the socio-economic and institutional milieu of the villages accounting for high proportion of suicide deaths among cotton growers.

(ii) To identify the socio-economic background of the suicide victims, their interactions and linkages with institutional and non-institutional delivery systems dealing in farm inputs, credit and farm products at local level; also to focus on the individual community linkages which will have a bearing on the problem;

(iii) To assess the causes for suicides based on a study of victims' households, perception of the neighbours, kin-network and elderly in the village and to suggest measures to prevent their recurrence.

Data Base and Sample Design

The data base of the Study consists of both secondary and primary data. Primary data are collected from institutions and individuals at various levels in the area and from sample households chosen for the purpose. This, it is hoped, minimises time and money in the conduct of field work, and facilitates an in-depth understanding of the Study. The list of mandals, and villages from which sample households are drawn for the Study are shown in Table 2.1.

Table 2.1 : List of Mandals and Sample Villages and Number of Sample Households : Warangal and Guntur : 1997-98

Name of the Mandal	*Name of the Village*	*No.of Sample Households*
1. Duggondi	1. Mohammodpur	2
	2. Nachenapalli	1
	3. Thimmampet	1
	4. Kesavapur	1
	5. Narayana Thanda	1
	6. Bandampalli	2
2. Atmakur	1. Kamaram	2
	2. Peddapur	1
	3. Kogilwai	1
	4. Agrahampahad	1
	5. Penchicalpet	1
3. Hasanparthi	1. Bairanpalli	2
	2. Pegadapalli	2
	3. Vangavapahad	1
	4. Jayagiri	1
4. Sangam	1. Rajyanaik Tanda	1
	2. Ramachandrapur	1
	3. Katrapalli	1
	4. Chintapalli	1
5. Venkatapur	1. Papayyapalli	2
	2. Kesavapur	1
	3. Narasapur	4
6. Devaruppala	1. Kamareddigudem	1
	2. Seetarampuram	1
	3. Kadivendi	2
	4. Pedamaturu	1
Total	.	36

Name of the Mandal	*Name of the Village*	*No.of Sample Households*
Guntur		
1. Guntur	Pottur	1
2. Pedanandipadu	Pedanandipadu	1
3. Prathipadu	Prathipadu	2
	Pathamallyapalem	1
4. Machavaram	Cargireddypalem	1
	Chennayapalem	1
	Singarayapalem	1
5. Chilakaluripet	Kavur	1
6. Atchampet	Velpur	1
	Tallacheruvu	1
7. Edlapadu	Dintenapadu	1
	Turlapadu	1
8. Tadikonda	Panidaram	1
	Bejastapuram	3
	Mukkamala	1
9. Thullur	Inavolu	1
10. Krosuru	Anantavaram	1
11. Amaravathi	Lingapuram	1
12. Veldurthi	Veldurthi	1
13. Durgi	Kancharagunta	1
14. Medikonduru	Medikonduru	1
	Mandapadu	1
	Siripuram	1
	Paladugu	1
	Visadala	1
15. Dachepalli	Mutyalampadu	2
16. Macherla	Kambampadu	2
17. Piduguralla	Julakallu	1
18. Rentachintala	Rentachintala	1
19. Phirangipuram	Nudurupadu	1
20. Mangalagiri	Nidamarru	1
Total		36

Methodology

The causes for suicides are manifold in nature, some external to and some internal to the victims' households. As such, a meaningful understanding of the problem hinges on a careful analysis of the external factors that have a bearing on the economy and well-being of the individual *viz.*, rainfall, cotton yields, prices, public institutions dealing in credit, input and product markets at the district and village level. Data on these aspects are culled out from the published sources such as district and village statistics as well as through discussions with select officials and non-official sources at various levels.

To get at the socio-economic environment internal to the households having a bearing on the victims' decision to commit suicide, a detailed questionnaire was prepared. It contained questions to understand him in relation to his family and kin-network, his economic position, his interaction and linkages with public and private institutions in credit, input and product markets as well as those rendering technical know-how

Suicides of Cotton Growers in Andhra Pradesh : Emerging Perspectives from Media

In sharp contrast to the suicides among farmers in Prakasam and Guntur districts during 1987-88, suicides of cotton growers in Andhra Pradesh during 1997-98 have evoked much response from academics as well as media. An overview of the academic effort at unearthing the factors accounting for this human misery has led to emergence of alternative perspectives. Broadly they can be grouped into the following:

Liberalisation Perspective

This perspective traces the mass suicides of cotton growers in Andhra Pradesh as a fallout of liberalisation policies pursued since 1991. The principal exponents of this analogy are Vandana Shiva and Utsa Patnaik among others. Utsa contends that "the small and the middle farmers are highly vulnerable to risk or accidents, for domestically consumed cash crops which are increased greatly for an export crop. The grower can do nothing about the volatile global price for the export crop neither can nation state" (Utsa Patnaik, 1998). Liberalisation they argue, by facilitating free trade has aggravated the problem by encouraging small and marginal who will get lost in the ebb and tide of exports.

Sobhanadreeswara Rao and Abdul Noor Bhasha also subscribe to this perspective on the ground that liberalisation reduced the flow of Bank credit to the agricultural sector and thus drove farmers to money-lenders and input dealers. They consider it as one of the major factors for cotton growers' suicides (Sobhanadreeswara Rao, 1998, Bhasha, 1998).

Technological Perspective

The protagonists of this view are mostly agricultural scientists. They traced the human loss to aping the developed countries' technology, particularly pesticide use which not only failed to contain pests but also destroyed the microbes which act as a catalyst between the soil and the plant (Venugopala Rao, 1988).

Institutional Perspective

This perspective attributes the problem to inadequate and untimely institutional credit support and extension facilities (Venkateswarulu, 1998).

Exogenous Perspective

This traces the crop loss and consequent human loss to inadequate rainfall during sowing season and untimely downpours in picking season, resulting in double disadvantage to the cotton growers. Many considered this as one of the factors. This has aggravated the incidence of pests.

Human Problem Perspective

This perspective traces the problem to spurious seeds, fertilizers and pesticides supplied by private sources and failure of Governmental agencies in arresting these and providing necessary technical know-how on time. Broadly, this is similar to the institutional perspective (Jagannadham Naldu, 1998).

Perspectives Based on Spurious Inputs and Their Indiscriminate Use

Agronomists are the protagonists of this perspective and they argue that shift in monsoon patterns, possibly due to eco-imbalances, has aggravated the uncertainty in rainfall and consequent need for higher dosages of pesticides leading to their indiscriminate use (Sampath Rao, 1998, 42).

Social Perspective

In their exploratory study which reflected on the alleged causes of sulcides reported in the media, Parthasarathy and Shameem looked on various parameters like rainfall, yields and prices of cotton, they identified inaccessibility of institutional credit and consequent dependence on land-owner-money-lender-combines as a factor for suicides among cotton growers in Warangal district. They looked at the entire problem as due to the emerging changes in social milieu alienating the individual from family and society.

It must be noted that most of the above attributes are not new to Indian agriculture which is devoid of an integrated agricultural policy in the larger interests of the country and the farmer. The fact that deaths have occurred in the pre-liberalisation era among cotton growers in Guntur and Prakasam districts in 1987-88 and in suicides of 1997-98 among growers of pulses in Karnataka, do not lend credence to the liberalisation perspective. It is difficult to reflect on the social perspective *i.e.*, whether all this is due to social disintegration for want of quantifiable social variables to discern the facts. In view of the alternative perspectives content analysis of the media reports was undertaken to see whether it helps in throwing further light on the above perspectives.

The identified causes in the media reports are broadly grouped into four categories *viz.*, natural factors like rainfall, causes relating to credit and indebtedness, those relating to inadequacies and lapses of Government agencies, liberalisation and Governmental Policies.

Content analysis of the reported causes for suicides among cotton growers clearly showed that, problems of credit, indebtedness and related causes emerged as the single largest factors. Next in order of significance are the lapses of the Government agencies dealing in extension, maintaining quality seed, fertilisers, pesticides and marketing etc. Natural factors like inadequate and untimely rainfall also played its due share accounting for around 23 per cent of the total reported causes. Causes related to liberalisation and economic policy of the government accounted for only 7 per cent. Farmer related causes like higher use of pesticides and personal reasons accounted for a small percentage.

District Level Scenario : Field Evidence

A comparative study of district level data with respect to natural endowments like soil types, irrigation, rainfall, agrarian structure, institutional credit, private dealers in inputs and agricultural production plans have sharply brought into focus the factors accounting for suicides in Warangal and Guntur districts.

Soil Types

Black cotton soil, the ideal type for growing cotton under irrigated conditions, formed only 22 per cent of the area in Warangal as against 70 per cent in Guntur. During 1997-98, cotton accounted for around 48 per cent of the cropped area in Warangal as against 32 per cent in Guntur. This shows that around one-half of the area under cotton in Warangal was sown on relatively less suitable soils and is subjected to high risks under rainfed conditions.

Irrigation Sources

Irrigation of cotton fields during flowering and fruit-bearing seasons helps augmenting yields. Warangal has only 1.5 per cent of area under canal irrigation and is largely dependent on wells and tanks. These sources are highly undependable in years of inadequate rainfall as in the case of 1997-98. About one-half of the area in Warangal is under dire need of Irrigation facilities because of soil types and is more vulnerable to lower yields compared to Guntur.

Rainfall

During the six year period 1992-93 to 1997-98, Warangal which usually receives its maximum rainfall through South-West monsoon, received the lowest in 1997-98 *i.e.* 570.7 mm as against the normal of 866.6 mm. Further, rainfall is only 62, 50 and 55 per cent of the normal during the cotton sowing season June, July and August. The rainfall during the north-east monsoon period *i.e.* October-December is double the normal *i.e.* 208 mm as against a normal of 108.1 mm. Further, in February 1998 it is 31.8 mm as against the normal of 14.6 mm. Thus, the cotton growers of Warangal suffered the double disadvantage of inadequate rainfall during the sowing season and downpour during the picking season.

Monthly rainfall data for the five year period *i.e.* 1993-94 to 1997-98 show that while cotton growers of Guntur also faced a similar situation during sowing season, the problem is less severe during picking season.

Agrarian Structure

Data on operational holdings examined for two points of time *i.e.* 1990-91 and 1995-96 do not show any perceptible difference between the two districts either in terms of average operated area of marginal and small farmers, or in their relative ratio of share in area to share in holdings. The average operated area of marginal and small farmers is around Ha. 0.74 and Ha. 1.40 in both the districts. However, one notable difference between the two is relatively higher percentage of holdings in the size-groups of Ha. 0.01-0.50 in Guntur compared to Warangal. The average size of operated area is higher in Guntur (Ha.0.30) while it is Ha.0.20 in Warangal in 1995-96.

Institutional Credit

Data drawn from Annual Credit Plans of the two districts show a relatively higher level of short fall in relation to achievements in Warangal under crop loans except during 1996-97 and 1997-98. It also showed that cotton remained a low priority crop even during 1996-97 and 1997-98.

Crop-wise analysis of credit flow showed that during 1996-97 and 1997-98 the achievements for cotton in Guntur exceeded the targets both in terms of number of accounts and amounts disbursed. In Warangal these proportions are 40.11,48.15, 55.29 and 54.11 respectively for accounts and amounts disbursed during 1996-97 and 1997-98.

In respect of term loans, for minor irrigation, the achievements are not only far lower but declined further in Guntur during 1997-98 compared to 1996-97. In Warangal too, the achievements under minor irrigation fell short of the targets during 1996-97 and 1997-98 as well. However, the percentage of achievement in Warangal is higher compared to Guntur and increased during 1997-98 compared to 1996-97.

The target fixed as a percentage of estimated demand for cotton (based on scales of finance) for 1997-98 formed 8.2 and 40.59 per cent respectively in Warangal and Guntur. The achievements as a percentage of estimated demand formed only 4.48 per cent in Warangal while it is 45.70 per cent in Guntur, though the estimated demand per hectare is higher in Warangal Rs. 19,019 as against Rs. 18,525 in Guntur. This shows the allocations as well as achievements are too meagre for cotton in Warangal compared to Guntur. Further, the allocation does not seem to take into account the percentage of the area under cotton as well as the higher per hectare credit requirements for the crop.

Agricultural Production Plans (APP's)

An incisive perusal of the APP's for Kharif 1997-98 of the two districts showed that cotton has not received the attention it deserves. Though it accounted for 1/3rd of the cropped area in the two districts, cotton did not find place in the eight crops selected for supply of seed in both the districts. Even under Intensive Cotton Development Programme (ICDP) the strategy is limited to supply of sprayers and sprinklers on subsidy basis in Warangal. It also remained as outcast in the special training programmes envisaged by other wings of the Agricultural Department in Warangal.

The APP of Guntur is better placed in this respect in the sense, it contemplated on a variety of demonstration farms and training programmes for farmers and clearly spelt out the daily routine of its functionaries and prescribed that agricultural demonstrators should examine crops in the villages on every Tuesday, Wednesday and Thursday. It also planned for distribution of cotton seeds through mini-kits in 14 out of 21 mandals growing cotton. Thus, in terms of strategy to increase production and productivity, APP of Guntur stands on a different footing than that of Warangal.

Input-Delivery System

Private sources dominate the seed and pesticides supply relative to fertiliser supply both in Warangal and Guntur districts. The share of co-operatives and agros in the total outlets which formed around 15 and 12 per cent in respect of seeds and fertiliser is less than 2 per cent in case of pesticides in Warangal. In Guntur, the share of institutional agencies is less than 4, 12 and 0.50 in respect of seed, fertiliser and pesticides. The relative insignificance of institutional agencies, particularly in seed and pesticides exposed farmers to the mercy of the private dealers. In Warangal the low level of institutional credit has closed options to the farmers either to select the dealer or the product.

Market Prices for Cotton

Monthly data on minimum prices of cotton in Warangal and Guntur regulated market yards show a highly erratic behaviour for the period 1993-94 to 1997-98. The minimum prices paid for cotton are relatively higher in Guntur compared to Warangal for 1996-97 and 1997-98.

Field Evidence at Village-Level : Social Background

Social background of sample villages in terms of dominant caste groups showed that 85 per cent of Warangal villages consisted of mostly backward classes while 87 per cent of Guntur villages are inhabited mostly by forward castes.

Soil Types

Only 70 per cent of the sample villages had black cotton soil in Warangal as against 94 per cent of such villages in Guntur district. This shows that nearly 1/3rd of the Warangal villages with mixed black and red soils faced problem during 1997-98 due to inadequate rains.

Infrastructural Facilities

Information elicited from 27 villages in Warangal and 31 village in Guntur districts on distance of villages from various types of infrastrutural facilities *i.e.* bus, train, post and telegraph offices, public telephone and nearby town which affect access to communication and availability of Government functionaries to the farmers showed marked differences between the two. Nearly more than one-half of the Warangal villages are located at a distance of 10 km or above from telegraph and public telephone, nearby town, bank branch and market yards for sale of cotton while such villages are few in Guntur district.

Differences are noted in respect of availability of private agencies dispensing credit and farm inputs. While more than 50 per cent of the sample villages had the above facilities at their door steps in Guntur, only 25 per cent of Warangal villages had this facility Thus, villages in Warangal district are at a greater disadvantage for want of access to credit and input services at their door steps compared to Guntur.

Marketing Practices

Marketing practices of marginal and small farmers in purchase of farm inputs and marketing of farm output showed that in Warangal around 88 per cent of sample villages farmers buy inputs and sell cotton in the nearby town while only a small per centage of villages in Guntur buy inputs outside their village and only 6 per cent sell cotton outside their village. Thus, the cotton farmer of Warangal is at a disadvantage particularly in the marketing of cotton, as he loses his bargaining strength and gets exploited despite so called regulated markets and he is at odds to take back home his product even if he is not satisfied with ruling prices on that day. The cotton farmers in Guntur enjoy a relative freedom in this regard.

Interest Rates in Informal Credit Markets

In more than 60 per cent of the Warangal villages the usual rate of interest on credit sale of seed, fertiliser, pesticides, consumption and production credit is 36 per cent. In contrast, more than 2/3rd of the Guntur villages reported the usual rate of less than 12 per cent in respect of inputs and credit. However, some of the villages reported usual rate of interest on credit sales of pesticides in the range of 24 to 36 per cent.

Villagers' Perception of Public Institutions

High per cent of Warangal villagers reported that they have never seen the functionaries of agricultural extension, banks or MARKFED with the exception of those of Co-operatives. They considered the information provided by them as not useful. In contrast, a high percentage of Guntur villagers reported greater access, usefulness of information provided by the functionaries of the agricultural department. Similar differences also persisted in perceptions regrading usefulness of services provided by the functionaries of banks and MARKFED.

Social Regulation

Social pathologists opine that rates of suicide are likely to be high in societies with high as well as low levels of social regulation (Lester David 1997 and 1998). Social regulation in terms of adherence to social norms and controls, control of faction groups over services, control of caste over group members, community concern for others' misery and participation in social and religious functions is more in Warangal villages compared to Guntur.

Field Evidence at Household Level

Household level survey has drawn information on the victims' households (bereaved families) from the remaining members of the household, i.e., father, wife and in some cases from kin-network.

Classification of Farms

For purposes of analysis, the sample households (SHH's) are grouped under 4 categories : Landless, Marginal, Small and Others. This is done on the premise that access to institutions, public and private in rural society, largely centres round one's ownership status in land. However, the usual farm management procedure *viz.*, classification based on operational area is adopted for computing cost as well as farm business incomes etc. The distribution based on ownership and operational holdings is as follows:

Size-group (Ha.)	*Ownership of Land*	
	Warangal	*Guntur (1997-98)*
Landless	3	9
0.01-1.00	11	16
1.01-2.00	13	7
Above 2.00	9	4
Total	36	36

A majority of the sample households in both the districts belonged to marginal and small farmer categories. Compared to Warangal, the percentage of landless as well as marginal farmers are much higher in Guntur if one goes by ownership holding. However, as per operational holdings, the per centage of marginal and small holdings is much higher in Warangal compared to Guntur.

Social Background of Households

Backward castes in Warangal and Forward castes in Guntur accounted for 58 and 80 per cent of the sample households. Around 70 per cent of the victims in Warangal and 53 per cent in Guntur were below 40 years. The per centage of illiterate in Warangal (58.33) is much higher compared to Guntur. If one groups the mere literates along with illiterates, the per centage changes to 64 in Warangal and 75 in Guntur. Overall the per centage of educated among the victims is higher in Warangal compared to Guntur.

Victims and their Household Characteristics

Responses on nature of wife's disposition (ordinary or nagging), disposition to his neighbours (friendly, helpful or otherwise), weightage to family members in decision-making in financial matters, show Warangal is better placed than Guntur. However, only 30 per cent of SHH's of Warangal had grown up children in contrast to those of Guntur (78%).

Farm Household and Kin-Network

Greater interactions with kin-network coupled with practices of giving and receiving will help forthcoming of moral support in times of crisis and thus acts as a cushion against suicidogenic impulses. Interaction with paternal and maternal relatives, cognatic kin and other relatives show that the frequency of visits is more or less the same in the two districts. As expected, the general psychology of the farm households in both the districts is that they rendered disproportionate moral support to their paternal and maternal kin groups than they received in their hour of need. While the cognatic kin relationship is more or less commercial in Guntur it is more of giving than of taking in Warangal.

Membership in Local Institutions

About 56 and 94 per cent of SHH's in Warangal and Guntur respectively were outside the fold of co-operatives. Similarly, 86 and 100 per cent in Warangal and Guntur were outside the fold of caste associations. This is true of gossip groups as well as faction groups in both the districts.

Size-wise, there is an inverse relation between size of holdings and membership in co-operatives in Warangal. This is largely true of Guntur also. In respect of other organisations, no particular pattern is discernible based on size of holding.

Land Ownership/Leasing-in and Average Size of Farm

Details of sample households by ownership holding, leased-in area and average size of ownership holding show that the total area owned by the sample households in Warangal is almost double of their counterparts in Guntur. However, the total leased-in area in Guntur is nearly four times higher than that of Warangal. This is largely due to the high per centage of landless in Guntur taking to cotton compared to Warangal. This is clearly reflected in the average size of operated holding.

Similar details of SHH's by operational holdings show the share of leased in area to operated area is 23 per cent in Warangal as against 67 per cent in Guntur. There is no relationship between size of holding and leased-in area in both the districts. The average operated area for all households is Ha. 1.80 and Ha. 2.38 respectively in Warangal and Guntur.

Cropping Pattern

Cotton accounted for a lion's share in the cropping pattern of SHH's in both the districts. There is an inverse relationship between size of holding and share of area under cotton in Warangal. The area under cotton is more than 80 per cent in the case of marginal and small farms as against 20 per cent in the case of large farms in Warangal. In Guntur, with the exception of the first size group there is a positive association between the size of farm and the share of area under cotton. This shows that the preference for cotton among the marginal and small farmers in Warangal is much higher compared to their counterparts in Guntur. This also partly explains the higher rates of suicides among this group in Warangal.

Assets and Liabilities

Structure of assets of SHH showed that land accounted for 80 and 64 per cent respectively in Warangal and Guntur districts. Next to land, dwelling house accounted for 1/4th of the total assets in Guntur as against 15 per cent in Warangal. The share of farm equipment in total assets formed 4 and 3 per cent respectively in Warangal and Guntur. Gold and other household durables formed only 0.75 per cent in Warangal as against 8 per cent in Guntur.

The average value of assets per household is Rs.2.29 lakhs in Warangal and Rs.1.20 lakhs in Guntur. These differences are largely due to the higher component of irrigated land and higher land values in the former compared to the latter. The average accumulated debt of these households (principle + interest on past loans) formed Rs. 1.21 lakhs and Rs.0.97 lakh per household. Debts as a proportion of assets formed 53 and 81 per cent in Warangal and Guntur.

Across size groups, there is an inverse relation between size of operational holding and proportion of debts to assets. Another notable feature is, an alarming disproportion between debt in relation to assets among the lower size-groups of holdings. This could be one of the factors motivating suicidogenic impulses.

Input Market and Farm Household

More than 80 per cent of the sample households purchased seed, fertiliser and pesticides from private sources in both the districts.

Current Borrowings

The situation in current borrowings for 1997-98 is similar to the one emerged in the case of outstanding loans. To illustrate, only two per cent of SHH's borrowed from commercial banks in Warangal. However, in Guntur around 17 per cent of the loans are from institutional agencies. Among the institutional sources, the share of commercial banks is much higher compared to co-operatives and RRB's in particular.

Among the non-institutional sources, though big farmers continued to account for a high proportion of loans in Warangal, commission agents emerged as the next dominant source accounting for 1/5th of the total number of loans. In contrast to Warangal, in Guntur input dealers emerged as a dominant source accounting for around 44 per cent of the loans taken during 1997-98.

Compared to share in number of loans, the share of institutional agencies in the current borrowings of SHH's is much lower *i.e.* 1.5 per cent and less than 10 per cent in Warangal and Guntur respectively. Among the non-institutional agencies big farmers and commission agents together accounted for around 60 per cent of the current borrowings in Warangal. Input dealers alone accounted for more than 60 per cent of current borrowings in Guntur district.

Distribution of Number of Loans and Outstanding Debt by Cash and Kind

Around 90 per cent of loans and outstanding debt during 1990-91 to 1996-97 in Warangal is in the form of cash. The proportion of kind loans as well amount taken in kind are more or less same in Warangal district. Size-wise there is an inverse relation between proportion of loans as well as outstanding credit dues and size of holding.

In Guntur, cash loans and outstanding borrowings formed

around 77 per cent. There are no observable difference either in the proportion of loans or outstanding amounts taken on credit. In contrast to Warangal, in Guntur there is a positive association between proportion of loans and outstanding credit.

Distribution of Loans and Current Borrowings by Cash and Kind

The pattern obtained in current borrowings for 1997-98 in Warangal is similar to past loans and outstanding dues in all respects. The only difference is a marginal decline in proportion of loans and amounts borrowed on credit during 1997-98 compared to the past.

Guntur presents a different picture with sharp decline in proportion of cash loans (amount borrowed) during 1997-98 compared to past loans and outstanding dues. For instance, the proportion of credit (kind) in total borrowing during 1997-98 is around 65 per cent as against 22 per cent in the past outstandings. This is in tune with the predominance of loans taken from input dealers in Guntur during the year under study. Leaving the first group, there is a positive association between the proportion of loans as well as amount borrowed on credit and an inverse relation between cash transactions and size of holding.

Rate of Interest

Around 75 per cent of the outstanding loans in Warangal are in the interest range of 36 per cent. Another notable feature is an inverse relationship between size of farm and proportion of loans contracted at interest rate of 36 per cent. In contrast to Warangal, around 60 per cent of loans are in the interest range of 24 per cent and around one-fourth at 12 per cent in Guntur.

In terms of amounts, around 70 per cent of the outstanding dues in Warangal are in the interest range of 36 per cent and above and 20 per cent in the range of 24 per cent. In Guntur only 14 per cent of outstanding dues are in the 36 per cent and above interest range, while around 70 per cent of outstanding dues are at 24 per cent. One feature common to both the districts is that landless and marginal farmers contracted loans at relatively higher rates of interests, compared to others.

With respect to current borrowings also a more or less similar pattern emerged. To illustrate, more than 80 per cent of the loans borrowed during 1997-98 and an equal proportion of the amount borrowed in Warangal carried an interest rate of 36 per cent. In Guntur 60 per cent of the loans and 71 per cent of the amount borrowed is in the interest range of 24 per cent. Another notable difference between the two districts is access to credit at relatively lower interest rates to SHH's in Guntur in contrast to those in Warangal.

Distribution of Loans and Outstanding Dues by Type of Security

More than 90 per cent of the loans as well as outstanding debt of SHH's in Warangal is on promissory note. In Guntur 60 per cent of loans and 70 per cent of outstanding dues are on the similar basis. Next to promissory note, land is the important source of security offered in around 1/4th of the loans and for 17 per cent of outstanding debt. The proportion of loans as well as amounts borrowed without any security formed around 10 per cent in Guntur while these figures are 2.5 and 5.28 respectively for number of loans and amount in Warangal.

In respect of current borrowings more than 80 per cent of loans and amount is borrowed on promissory notes. The per centage of loans and amount taken without any surety formed 13 and 17 per cent respectively in Warangal. In Guntur the situation is not only different from that of Warangal but also in relation to the pattern noted in respect of past-loans and debts security wise. There is a significant rise in the percentage of loans and amounts borrowed without any security (*i.e.* 40 per cent of loans and 66 per cent of amount), only 36 per cent of loans and 21 per cent of total amount is borrowed on the basis of promissory notes. Land as a source of security formed only in 10 per cent of loans and around 5 per cent of the amount was borrowed showing land as security during 1997-98. Security-wise, there are no differences among size-groups, nor there is any pattern in the two districts.

Purpose-wise Borrowings

Purpose-wise analysis of past loans and outstanding dues show that 80 per cent of loans and 74 per cent of outstanding debts in Warangal are incurred for purchased inputs only. This holds good across all size-groups though, around 7 per cent the past loans are borrowed for consumption purpose, in terms of amount they constituted around 4 per cent only

Guntur presents a different picture. For all SHH's taken together, 47 per cent of past loans and 45 per cent of outstanding dues are incurred for purchased inputs. Next to inputs, loans for consumption purposes and for payment of rent formed 24 and 20 per cent respectively. However, their share in outstanding dues is only 26 for consumption and 18 per cent for payment of rent. There is no relationship between size of holding and the per cent of outstanding dues for buying inputs and a positive association between per cent of dues towards payment of land rent in Guntur.

In respect of current borrowings also, purchased inputs accounted for a major share both in respect of loans as well as outstanding dues in both the districts. However, these proportions are relatively lower in Guntur compared to Warangal. In Guntur, around 40 per cent of the loans and 14 per cent of the outstanding dues accounted for payment of land rent. In this respect, there is a positive association between size of holding and the per centage of loans and amounts borrowed for payment of rent. Around 6 per cent of the current borrowings are for consumption purpose in Guntur as against less than 2 per cent only in Warangal.

Linkages between Credit and Other Markets

Responses of the SHH's on the linkage between credit and other markets, both in respect of past borrowings and current borrowings are shown separately. In respect of the past loans and outstanding dues, more than 80 per cent of the SHH's in Warangal reported absence of any linkage between credit market and product market. Sale of product at market price was reported in 17 per cent of the SHH's in Warangal. However, in the case of landless one-half of the SHH's reported that there is an agreement to sell the product to the credit agency but at the ruling market price.

The proportion of SHH's reporting linkages is on a higher level of 60 per cent in Guntur in contrast to Warangal. Another notable feature in Guntur is forward trading (*i.e.* sale of crop at pre-agreed price) and repayment in cotton among landless and marginal farmers though their proportion is not very high.

In respect of current borrowings, the percentage reporting linkages formed 25 per cent in Warangal and 36 per cent in Guntur. Further, the payment in cotton is reported by around 8 per cent of the landless and 5 per cent of the marginal farmers in Guntur.

Source of Technical Knowledge about Cotton

Nobody in Warangal reported agricultural extension staff as a source of technical know how. About 94 per cent reported big farmers as their guides. But in Guntur, 44 per cent reported agricultural extension staff, and another 30 per cent reported input dealer as source of knowledge in matters relating to agronomic practices.

Gross Income and Farm Business Income per Hectare of Cropped Area

Gross income per hectare is higher in Guntur compared to Warangal by about Rs.2000. Due to the unique situation of 1997-98, Farm business income, Family labour income and Net income from cotton turned out to be negative in both the districts. The per hectare figures are much higher in Guntur compared to Warangal.

Expected and Actual Yields and Income from Cotton

As mentioned earlier, it is a mis-match between expectations and achievements that nurtures suicidogenic impulses. Higher the mis-match over time greater the scope for temptation towards suicide. Despite the ups and downs of farming in general and cotton in particular, the increase in area under cotton within the districts as well as in the villages which are mute witnesses to suicides among fellow farmers need reflection. Under unirrigated conditions there is no other crop with the exception of chillies that ensures higher yields under normal conditions. The price risks in chillies, not yet covered by minimum support prices is another reason for continued growing of cotton against all odds.

The expected yield of cotton per hectare of cropped area is around 22 quintals in both the districts. Size-wise, these are slightly higher among the lower strata of farmers. The realised yield per hectare during 1997-98 is 9 and 8 quintals in Warangal and Guntur respectively. The actuals as a per centage of expected yields in Guntur are in the range of 28 to 35 per cent as against 30 to 45 per cent in Warangal. There is a positive relationship between size and actual yield as a proportion of expected yield and income in Warangal. No such trend is noted in Guntur

It is the mis-match between expectations and achievements that nutured suicidogenic impulses. Higher the mis-match as in 1997-98, greater the scope for temptation to commit suicides.

Timing of Sale of Cotton

The timing of sale of cotton is also crucial in getting remunerative prices. Though, difficult to predict, it is common knowledge that post-harvest prices are likely to be lower due to distress sale among farmers, high moisture content and wait and see policy of the traders.

Around 61 per cent of SHH's of Warangal sold away 90 per cent of cotton produced during 1997-98 in December 1997 at a minimum price Rs. 1,700 per quintal. Only 10 per cent of them sold less than 10 per cent of their output during January and February when prices increase. Cotton prices increased by Rs. 400 to Rs. 500 per quintal in January and February 1998 compared to December 1997.

In contrast to Warangal, in Guntur only 14 per cent of SHH's sold around 1/3rd of their output in December 1997 and 87 per cent sold 71 per cent of their output at the highest prices of more than Rs.2,000 per quintal, during January-March 1998. This shows cotton gowers of Warangal have also suffered through early sale at lower prices.

Another feature is near distress sale in December 1977 among all except in the case of marginal in Warangal. In contrast, in Guntur broadly there is an inverse relationship between size of holdings and quantity sold during December 1997 when prices are at the low ebb and positive association during February 1998 when prices are high. This reflects on the poor withholding capacity of the marginal and small households, and the resultant lower prices received by them in the product market.

Distress Sale of Cotton

Nearly 86 per cent of the SHH's of Warangal cited compelling family needs for early disposal of cotton soon after the harvest, while about 52 per cent of them in Guntur reported repayment of past debts as the reason for distress sale of cotton.

With regard to market channels, in Warangal 94 per cent of SHH's sold cotton through private traders, whereas in Guntur SHH's sold through both private dealers and input dealers. Size-wise, there is a positive association between size of holding and proportion of sellers through private traders in Warangal and in case of Guntur no discernible pattern is noted.

Sale of Cotton for Cash /Credit

Usually farmers sell cotton both for immediate/down payment and also on credit basis for 15-30 days. Price received on cash sales is often lower than the credit sales. Payment-wise, 72 per cent in Warangal and 90 per cent of farmers in Guntur sold for cash, while, the remaining sold on credit for payment within 15 to 30 days. In this connection, it is noted that in Warangal a farmer has to lose 3 per cent of the sale value if he opts for ready cash. No such deduction is noted in Guntur though, marginal differences persist in the unit sale price for ready cash and for sale on credit for a month.

Causes Leading to Suicides

Information was elicited from SHH's on the hypothesised causes that triggered impulses for suicides such as (1) loss of crop, (2) feeling of economic bankruptcy, (3) loss of face, (4) lack of moral support from family, friends, and/or kinship group, (5) worry about future; and, (6) harassment for repayment of debts. Feeling of economic bankruptcy and harassment for repayment of debt are reported to be the major causes for the suicides. However, the per centage reporting economic bankruptcy is more in Warangal compared to Guntur.

Similarly, the per centage reporting harassment is more in Guntur than in Warangal. Only one person reported harassment in Warangal, while more than 3/4ths of those reporting harassment stated private creditors as the major source of harassment.

Conclusion

To sum up, none of the alternative perspectives individually or severally account for the causes of suicides among cotton growers in Andhra Pradesh. Their contribution lies in identifying the causes for the tragic losses on crop front. In this context it must be noted that the growers who are used to ups and downs in farming are unlikely to end themselves for a crop loss in one single year, however unbearable it may be. The macro and micro level evidence clearly affirm that the marginal and small farmers have entered the arena of high risk commercial farming in the hope of salvaging themselves. This they are venturing under an unenviable environment of inadequate resource endowment, inadequate and eroding institutional support for credit, inputs and marketing on one hand and unresponsive extensive networks and machinations of exploiters in all markets on the other hand. The cyclical nature of farming with occasional blessings kindered hopes of adventurism towards risk taking. When these expectations get belied either due to nature's fury, or human greed, or institutional indifference, life becomes vulnerable to the small and the weak. It becomes terrible if a combination of all the above occur similar to the situation of the cotton growers of Andhra Pradesh during 1997-98. Under the above circumstances which propel economic bankruptcy among the vulnerable, likely to trigger suicidogenic impulses in areas with high degree of social cohesion as well as those with low degree of social integration, in the former out of compulsions to adhere to social values and in the latter due to relative isolation and resultant alienation. Evidence from Warangal and Guntur represent the above scenarios. They also indicate that institutional support in its varied forms provision of services and know-how, helps a lot in diluting the intensity of the crisis while its absence aggravates the problem.

REFERENCES

Anjaneya Sastry, V. "Driven to suicide by Nature, Government". *The Hindu* 19th January, 1998.

Anonymous Writer , "Suicides by Farmers" (Letter to the Editor) *Economic and Political Weekly,* Vol.33 (II): p. 550, 14th March 1998.

Aron, Raymond, *Main Currents in Sociological Thought,* Penguin Books, Harmondsworth, England, 1967.

Bhasha, Abdul Noor, "Atmahatyala Viplavaniki Haritha Viplavarme Nandi', *Vaartha* (Telugu), 4th Feb, 1998.

Chalam,K.S. "Bhoo Poratalu-Atmahatyalu", *Vaartha*, 22nd April 1998.

CMIE, *India's Agricultural Seclor-A Compendium of Facts,* Center for Monitoring Indian Economy Pvt. Ltd., Bombay, 1955.

CMIE, *Agriculture,* August, 1997, Center for Monitoring Indian Economy Pvt. Ltd., Bombay, 1997.

Cohen Albert K., "The Sociology of the Deviant Act: Economic Theory and Beyond", *American Sociological Review* Vol. 30(l), pp.5-14. 1965.

Correspondent, "Uncertainty over Cotton cultivation", *The Hindu,* 29 th January, 1998.

Diwakar, "Atmahatyalu Kavu, Andolane Parishkara Margam". *Vaartha,* 17th February, 1998.

Editorial, "Cotton Deaths". *Business Line,* 27th January, 1998.

Editorial, "Death by Farming". *The Hindu,* 31st January, 1998.

Editorial, "Raitanganiki Tirigi Daga". *Vaartha,* l6th April, 1998.

Editorial, "Parapathi Sanskarana. Oka Paraswarme", *Eenadu,* 23rd April '98

Fertilizer Association of India, *Fertilizer Statistics,* Fertilizer Association of India, 1995-96.

Goldbratt Harold, *The Causes of Suicide,* Routledge and Kegan Paul, London. 1978.

Jairam Ramesh, "Pranalu Haristunna Patti" *India Today,* 19th May, 1998.

Joint Director of Agriculture, *Annual Production Plan, Kharif, Guntur District.* Joint Director of Agriculture, Guntur, 1997-98.

Khergamvala, F.J. "In this season for suicides, escapism can be a reason".*The Hindu,* 29th March, 1998.

Lal, M.B., "Dark Clouds in the Cotton World", *The Economic Times,* 26th January, 1998.

Lester, David, "Can there be a Durkheimian Theory of Homicide?, *Indian Journal of Criminology,* Vol.20(2), pp.81-85, 1992.

Naidu, Jagannadham P., "Palakula Padaganeedalo Vyavasaya Rangam' *Vaartha,* 15th February' 1998.

Naresh Kumar, S, "Ex-Gratia, no solace for Cotton Farmers", *The Hindu,* 10th January, 1998.

Naresh Kumar, S. "Where Cotton suffocates". *The Hindu,* 22nd February'98.
Narayana Reddy, B. "Bathuku Bharame Nannenduk, Batikincharu", *Vaartha,* 15th February, 1998.
National Crime Records Bureau, *Accidental Deaths and Suicide in India,* National Crime Records Bureau, Ministry of Home Affairs, Govt. of India, New Delhi, 1995.
Parthasarathy, G. & Shameem, "Suicides of Cotton Farmers in Andhra Pradesh —An Exploratory Study", *Economic and Political Weekly* 33(13), 720-26, 1998.
Patnaik, Utsa, "Food Security, Globalisation and Poverty towards the New Millennium" paper presented at the Plenary session 1 *of VIII National Conference* of *Women Studies on Survival and Sovereignty,* 1998.
Prasada Reddy, M.E.V., "Market ku Taloggutunna Raitu" *Vaartha,* 16th April, 1998.
Prasada Reddy M.E.V., "Guptha Nivedika Vyavasaya Parapathi Penchutunda?" *Vaartha,* 24th April, 1998
Parasanth Reddy, CH, "Little Guidance adds to A.P.Cotton ryots Woes" *Business Line,* 28th January, 1998.
Rajendra Prasad, R.J., "The A.P.Verdict".The *Hindu,* 18th March, 1998.
Raju, N. N., "Atmahatyalu-Sastreeya Dhrukpadharn" *Vaartha,* 19th February, 1998.
Ramesh Chand and Birthal Pratap, S., "Pesticides use in Indian Agriculture in relation to Growth in Area and Production and Technological Change", *Indian Journal of Agricultural Economics,* Vol. 3. July-Sept., pp. 488-498, 1997.
Ramoo, S.K., "Cruel Nature, Lack of guidance drives them to suicide". *The Hindu,* 2nd March, 1998.
Rao, V.M., "Farmers in Market Economy, Would Farmers gain through Liberalisation?" *Indian Journal of Agricultural Economics,* Vol. 49. No. 3, July-Sept. pp. 393-402, 1994.
Reporter, "174 ku Cherukunna Raitula Atmahatyalu". *Vaartha,* 16th February, 1998
Reporter, "Mass hysteria valle Raitula Atmahatyalu". *Eenadu,* 9th March, 1998
Reporter, "Record Sthayilo NABARD Re-Finance", *Eenadu,* 2nd April, 1998.
Reporter, "Kendra, Rashtra Aardhika Vidanale Raitula Atmahatyalaku Karanam" *Eenadu,* 10th April, 1998.
Reporter, "Raitula Adukone Theeru Idena?" *Eenadu,* 20th May, 1998.
Reserve Bank of India, *Report on Currency and Finance* 1995-96, Vol. 1, *Economic Review,* Bombay, 1997.
Sampath Rao, S., "Sagatu Raitu Samasyalaku Parishkaramedi", *Annadata,* July, pp. 42-43.
Sarath Babu,B., "On the suicides of cotton farmers in A.P." *The Hindu,* 19th February, 1998.
Shiva Vandana & Afsar H. Jafri, "Seeds of Suicide, The Ecological & Human costs of Globilisation of Agriculture", Research Foundation for Science, Ordinary Technology and Ecology, New Delhi, 1998.
Shobanadreeswara Rao,V., "Annadatalaku Atma Viswasarm Kalpinche Charyalu Mukhyam". *Eenadu* (Telugu), 20th March, 1998.

Staff Correspondent "Alarm over 'suicide' by Punjab Farmers". *The Hindu,* 28th February, 1998.

Sudhakara Reddy, B.K., "The lessons we refuse to learn from cotton deaths".*The Economic Times,* 17th January, 1998.

Sudhakara Reddy, B.K., "Brown Spot Disease affects FCV Tobacco crop in *A.P":The Economic Times,* 18th January, 1998.

Venkateswarlu, V., "Annadatala Atmarpanalaku Antham leda?" *Vaartha,* 23rd April, 1998.

Venkateswarlu, V., "Appulu, Vaddeela Vishavalayamlo Raitulu Vilavila", *Eenadu* (Telugu), 19th May, 1998.

Venugopal, N., "Were Suicide Deaths of Farmers Avoidable" ? *The Economic Times,* 1st January, 1998.

Vanugopala Rao, N., "Patti Raitunu Patti Peedistunna Rajakeeya Cheeda Peedalu". *Vaartha,*19th February, 1998.

Vidhyabhushan, "Aspasta Vidhanale Atmahatyalaku Karanam". *Vaartha,* 27th February 1998.

3

State of Cotton Farmers in Andhra Pradesh

NABARD Chair Unit

The distinct transformation one can witness in State agriculture is undoubtedly growing commercialisation of farm sector. Uptrend in area under cash crops, pursuance of hi-tech farm practices and larger monetisation are evidences of this metamorphosis in our economy. Apart from the surging demand world-wide, the scope for increased returns brought about new dimensions in commercial agriculture. Following the introduction of high-yielding varieties, the emphasis has been on intensification so as to enhance productivity. This approach bestowed scant regard to sustainability and resulted in new generation problems in commercial crops like cotton.

One will not fail to observe the capital intensity in hi-tech cotton farming. Large outlay is necessitated on account of pest surveillance and control at every stage of the plant life. The inputs cost is about Rs. 10,000 per acre. The cotton farmers in Telangana are mostly those migrated from neighbouring districts like Prakasam and Guntur take land on lease for growing this crop. Being cultivators of modest means, the lease money of Rs.3,000 to Rs.4,000 per acre adds to their financial burden as also the high interest outgo on borrowed funds. They might be approaching the money lenders for financial assistance as banks will not be in a position to lend to lessees. Even in normal course, bank assistance is restricted on the basis of scale of finance, availability of lendable resource etc., and the average loan amount for cotton crop seldom crosses Rs.3,000 per acre.

Certainly a thorny environment of cotton growing in the State, Telangana region in particular. The unprecedented crop failure last year aggravated their miseries and rocked up their very survival. The yield was bad and the market was depressed, cotton farmers underwent mental torture and there was public outcry.

The weaker sections, *i.e.* small and marginal farmers suffered more. They were attracted to cotton cultivation in the light of fortunes reported in the neighbouring States of Maharashtra and Karnataka. For them, in the context of greater cash needs, switching over to commercial agriculture is sheer necessity with the constraints of land and limited possibility of Rabi crop due to water scarcity even under irrigation source. The small farmers naturally became optimistic and diverted all their resources, both owned and borrowed, to cotton growing, hoping that the market mechanism and State machinery would extend necessary support and not play foul. The farmers were late to note that spurious seeds, fertilisers and pesticides were freely let into the market with their respective sales personnel marketing the products in high pitch. The price of cotton which was Rs. 2,200 per quintal in early 1997 crashed to Rs. 1,600 when the Kharif crop arrived at the market in late 1997/early 1998.

Infrastructural Deficiencies

The above analysis exposes the infrastructural deficiencies rather than farmers inefficiency in explaining the suicides we are now concerned with. Development of suitable rural infrastructure is crucial for the growth of commercial agriculture. In the case of cotton, the access to quality seeds, fertilisers and pesticides would have been ensured through highly committed, service-oriented distribution net-work. The quality control units should have been in place to check the infiltration of spurious inputs. The Agriculture Department should have endeavoured to disseminate necessary knowledge to the prospective cotton growers on different aspects of hi-tech cultivation practices well before they embark on the same. Extension work must be based on the needs and interests of the cotton farmers. These needs differ from farmer to farmer and from village to village. Good extension work should address towards assisting rural families to work-out their own problems and findout necessary technical know-how and arranging backward and forward linkages. This means that going by past record, a great leeway has to be made in restructuring extension force so that in the hour of crisis, the government machinery should be able to tackle the situation effectively without leaving the farmers in the lurch. Demonstrations, group discussions, campaigns, exhibitions, etc., should be more intensively organised along with ensuring adequate supply of inputs so that farmers will be fully equipped to meet any eventualities.

Hi-tech agriculture is a risky proposition in so far as the infrastructural bottlenecks galore coupled with weak crisis management mechanism. In the backdrop of the suicides, it is doubtful whether private efforts in commercial agriculture would come forth unless serious efforts are made in ensuring the enabling infrastructure—water through irrigation works, power supply for lifting water, round the clock pest surveillance and follow-up, quality control of fertilisers, seeds, pesticides, etc. besides a mechanism for continuous flow of scientific information on cotton cultivation to the needy.

Micro Planning

Increasing the yields from small operational holdings is often affected by constraints on the availability of land and water resources besides managerial weaknesses. The problems affecting cotton crop differ from region to region and within region from one locality to another. While there is a crucial need for a macro policy for creating a favourable environment for better flow of information, investments, inputs and technology, considering the fact that there are large number of actors managing the farm operations with wide variations in their resource position and risk bearing capacity, there is a dire need to micro planning/ decentralised planning if production is to be enhanced village by village. The situation calls for location-specific plans on the basis of local characteristics. The benefits can be conceived as creating durable social and economic assets for sustainable income generation from cotton.

Building up Managerial Skill

It should be emphasised that a technology driven package can be successfully grounded only when the farming community is competently equipped in managerial skills. The cotton crop being both technology intensive and capital intensive, organisation of cotton production should be treated as a highly skilled and risky job. The starting point is to expose farmers to new vistas of not only crop management but capital management as well. Cotton cultivation has to be made more cost-effective, competitive, sustainable and profitable with a comparative edge over other crops. In this endeavour, integrated nutrient, water and soil management systems have to be incorporated in the technology packages. The spate of suicides by cotton farmers is a clear indication of their dependency arising from lack of such managerial capacity to withstand crop failures. Thus, there is an immediate need to address to this problem as to how best we can inject specialised skills to our farmers to evolve farm plans, sequencing/orchestration of various operations including credit availment and marketing. Transfer of farm-worthy technology is vital for harnessing the fruits of research. In this evdeavour, re-orienting of the agriculture extension system to make it technology-oriented and demand-driven with appropriate policy setting, extension planning, technology assessment, refinement and transfer through innovative methods that help extension system to respond to the changing diverse needs of different agro-climatic situations would be important. Special emphasis will have to be laid on the extension needs of hilly, tribal and rainfed areas where cotton is grown. Transfer of farm worthy technology with impact analysis to ensure equity, social justice and overall growth would be the key to success. Perhaps, highly qualified extension staff may be required for the purpose to enable continuous interface between technological improvements and farmers adoption practices on cost-effective lines. Safety of natural resources is also an area of concern and therefore an element of environmental protection, resource conservation and its rational utilisation add challenging dimensions to research and extension processes in cotton development.

Development Banking

Agricultural development process involves several inputs and functions (irrigation, seed, fertiliser, pesticides, credit, marketing, know-how, management etc.) and interestingly illiterate, resource-poor farmer is expected to procure and combine them to earn reasonable income. In the absence of organised efforts, a farmer willing to take up commercial agriculture is bound to fail miserably for obvious reasons. The Nationalisation of Commercial Banks in 1969 and increased thrust on agricultural lending thereafter through multi-agency approach paved way for development banking in rural sector. Technically qualified staff now manage the agricultural lending operation in most of the rural branches. Though this has improved the quality of credit appraisal, there is still a large scope for bringing all the cultivators under development banking concept. At NABARD level, the importance of creating awareness among the peasants led to promoting VVV programme through farmers clubs all over the State. Infrastructural development is attempted through continuous dialogue with line departments and following it up in Potential Linked Credit Plan for each district in the State. Generous funding of irrigation projects is now taken up under Rural Infrastructure Development Scheme. Human resources development in banks is vigorously pursued through offering training facilities to bank staff. NABARD also provides financial support to co-operative banks and Regional Rural Banks for setting up technical, monitoring and evaluation cells for upgrading banks' capabilities in the areas of credit scheme formulation, appraisal and monitoring. Above all, it has embarked upon evolving Development Action Plans by signing MoU with banks to improve their lendable resources and to ultimately wean away farmers from the clutches of money lenders.

Special Focus on Small/Tenant Farmers

Adversity in farm front affects the small farmers and tenant farmers most in view of their weak staying power. Farm prosperity will be a myth if such weaker sections who form the vast majority are not cared for. NABARD has formulated a series of measures for strengthening their economic base. Significant among these are :

(i) Fixing norms that atleast 20% of the crop loans be disbursed for SF/MF.

(ii) Atleast 60% of the investment credit to be disbursed for SF/MF.

(iii) Granting of a separate line of credit to the co-operatives to cater exclusively to the needs of tribal farmers for an aggregate amount of Rs.400 crores as envisaged in the Union Budget 1995-96.

(iv) Separate line of credit for financing the SC/ST beneficiaries to the extent of Rs. 100 crores for investment credit purposes.

(v) Setting up Small Farmers Agri-Business Consortium to ensure proper forward and backward linkages with emphasis on value addition.

(vi) Extension of 100% refinance facilities to banks for providing credit support to SC/ST credit plans.

(vii) Provision of refinance assistance to poverty alleviation programmes and employment generation programmes such as IRDR

(viii) The GOI has also been providing substantial subsidy on sprinkler and drip irrigation systems which has been recently enhanced to 90% of the cost of system for SF/MF, SC/ST beneficiaries, etc. Assistance from NABARD is also extended in respect of such investments.

Crisis Management

On the crisis management front, measures like crop insurance and development of non-farm sector stand out. Writing-off loans cannot be advocated as it has unpleasant backlash effects on the banking ethics.

(i) Crop insurance scheme has to be made purposeful and farmer friendly so that even for minor loss, he has to be compensated fully and quickly. This is necessary to sustain interest in commercial agriculture especially in risky ecologues of Telangana. We may also have to identify other occupations in rural areas as a 'life-saving' device.

(ii) Non-farm sector would fit well to support such farmers wrecked by crop losses. For this purpose potential areas in non-farm sector which can yield reasonable returns may have to be identified and promoted. NABARD has pioneered in this field and its District Credit Plans do suggest promising areas for adoption.

(iii) At present cotton development is taken up in such a manner where accountability is non-existent. There is need for a single agency in the State which can provide specialised services ranging from assessing demand-supply conditions, training, seed supply to marketing, market intelligence and insurance. The financing arm of this outfit should be able to meet crop loan requirements in full. There is no reason why a total management system like this (as against the present fragmented system) cannot be extended to small peasant cotton farmers. Tea Board, Rubber Board and Coffee Board etc. are doing excellent services in their respective commodity groups.

(iv) When the State capacities are weak both for cotton development and for combating a crisis, it should have clear strategies well before the peasants take up the cultivation of this commercial crop. Contracting out certain services to private secrtor and NGOs can be thought of after adopting a carefully designed sectoral approach and a strong monitoring mechanism. Where feasible, direct participation of beneficiaries in the design, implementation and monitoring should be encouraged. NGOs should be allowed to function in critical areas where the record of Government machinery is not promising. Creating awareness among farmers on their rights as consumers (of farm inputs), and on the latest cost saving, improved technologies besides arranging testing of seed, soil, fertilisers, pesticides etc. can be attempted by NGOs. In fact, NABARD has already engaged an NGO for the purpose in Khammam District. The Krishi Vigyan Kendras also need restructuring for taking up similar tasks in cotton belts of Andhra Pradesh.

(v) Considering the complexity of extension services required for a crop like cotton, accredited private agencies may also be encouraged to provide specialised extension services on ongoing basis. In the context of rewarding and assuring results, cotton farmers will be prepared to bear the cost of the services offered by such private agencies.

(vi) In order to propagate sense of self-help and organised efforts in adoption of scientific farming techniques with the support of credit, VVV clubs have greater relevance. Each rural bank branch should promote a VVV club to bring about rural prosperity on sustained basis and forge a link between credit and commerical farming to the highest degree. It is also necessary that such VVV clubs are federated at mandal/district/State level to form an effective voluntary organisation which can protect the interests of the farmers in general and those farmers engaged in commercial agriculture in particular.

(vii) There should be greater resolve to check the distribution of spurious seeds, fertilisers, and pesticides through setting up of well equipped quality control laboratories in each mandal of cotton growing tracts. Frequent inspections and raids on dealers outfits is necessary to instill confidence among cotton growers.

(viii) The significance of primary sector and for that matter commercial agriculture can not be ignored any more. The terms of trade for agriculture, commercial crops in particular, should be favourable enough to attract more investments. The support price for cotton has to be remunerative not only to cover the cost of cultivation but to leave attractive profit to the farmer. Sectoral planning need to be attempted through precise forecasting the demand supply scenario and its impact on farm profitability. Farm planning can be designed in advance taking into consideration these market trends and price movements. There is also a need for healthy relation between farm gate price and price of finished product (cloth) so that the cotton farmer realises larger share of consumer rupee. Distress sales should be arrested by providing warehouse facilities and future trading. The lot of these farmers can be improved further if value adding post-harvest technologies (for example ginning facility) are available to them at their door steps.

(ix) One cannot address the problems of cotton farmers in isolation of social issues. Provision of primary education, regulation of tenancy, ceiling on and redistribution of land etc., may go a long way to reduce social tensions in cotton tracts. Systematic land records will have to be maintained at village level besides facilitating the poor tenants to beget the right to own the land in due course.

4

Why the Suicides and What is the Crisis?

A Case of Andhra Pradesh

A. PRASADA RAO

Present crisis in agriculture sector, in our country, is not totally unanticipated ever since liberalisation economic policies were sought to be introduced. Many social scientists including noted economists, have been warning the nation of such impending crisis. What has shocked everyone of conscience is the manner and severity with which the crisis has manifested, specially among the farmers in Andhra Pradesh. Though the farmers committing suicide were raising several crops, suicides of cotton farmers were highlighted in the press. These suicides, as seen from paper reports, also took place in Maharashtra, Punjab, Karnataka, Orissa etc. However, suicides were relatively higher in Andhra Pradesh, as compared to any other State.

According to Development Research Advisory Group of AWARE (Madhavan *et al.*, 1998), 89 per cent of the deceased/ sample farmers in Andhra Pradesh (92 samples out of 307 suicides between October, 1997 and April, 1998) were growing cotton, 43 per cent rice, 18 per cent sorghum, 17 per cent chillies, 15 per cent pulses and 13 per cent groundnut. The total comes to about 195 per cent. Evidently the deceased were growing more than one crop simultaneously. Hence, recent farmers' suicides should not be viewed as those of cotton farmers alone but as a general manifestation of crisis in agricultural sector. Chief Minister of Andhra Pradesh, while responding to the debate, in the State

Legislature to the charge of opposition against his Government being responsible to the then suicides of some farmers (Rabi of 1996-97) due to failure of the Government to ensure power supply to agriculture sector in March, 1997, also stated that it was the entire gamut of liberalisation policies that were being followed by the Centre since early 1990's, as being responsible for the suicides and not due to failure of his Government to ensure enough power supply to agriculture. At about the same time, Andhra Pradesh Rythu Sangam launched a State wide campaign "Rythu Rakshana Yatra", with the main theme to "Save the Farmers". During the campaign, the issues that came to the fore were declining support of State and Central Governments to agriculture, failure to ensure remunerative prices to the produce of farmers, protection of farmers in the Market Yards from deceitful and exploitative traders/brokers, and ensuring enough power supply to agriculture. Opposition to the then proposed restructuring of APSEB, which has been subsidising power for irrigation, specially in chronically drought affected areas in an exemplary manner, also became one of the focal point during the above agitation. Following continuing suicides, Government of India have appointed a High Power Committee, at the request of Government of Andhra Pradesh, under the Chairmanship of Sri. Bhagath Singh, I.A.S., Additional Secretary to Government of India, Department of Agriculture and Cooperation, to study the problems relating to prevailing distress among the farmers owing to indebtedness. From the above, it becomes clear that distress conditions have been prevailing among the farmers since a fairly long time due to one reason or the other and the decision of farmers to commit suicide in the recent past is neither a sudden emotional outburst or due to failure of cotton crop following aberrant weather conditions, and uncontrollable pest problems during 1997-98 crop season.

The Farmers Worst Affected

Deceased and sampled farmers cultivating less than 3 acres land, either as owner or tenant, constituted 50 per cent in cotton, 85 per cent in rice, 88 per cent in sorghum 94 per cent in chillies, and 88 per cent in pulses. No farmer falling under this category was growing groundnut. Similarly, farmers cultivating between 3 and 6 acres, constituted 44 per cent in cotton, 83 per cent in groundnut, 15 per cent in rice, 6 per cent in chillies and 7 per cent in pulses. Farmers cultivating between 7 and 9 acres, constituted 7 per cent in cotton 17 per cent in groundnut and 6 per cent in sorghum (Madhavan *et. al*, 1998). It implied that it was the small farmers, who were the worst affected, cultivating both commercial and food crops simultaneously

Did Market Forces do any Justice to Small and Marginal Farmers?

Opinion on effect of seeds, fertilisers and pesticides were sought from the families of deceased farmers in the above study. About 90 per cent of the samples, have opined that the fertilisers they got were either good or normal. Only 8 per cent of them felt that they were adulterated. Contrary to this 62 per cent of those sampled felt that the pesticides they got were adulterated. Similarly, 29 per cent felt that the seeds they got were adulterated (Substandard or spurious). If the fact that the manufacture and distribution of the bulk of fertilisers, barring those containing micro-nutrients are under the control and supervision of the State (broader sense), the public sector role in case of pesticides, and role of public institutions still being dominant, atleast in case of varieties despite encouragement of private seed traders under the New Seed Policy, are kept in view, the above data suggest that 'market forces with private trade' can not be depended upon to take care of the interest of the small and marginal farmers, who constituted the bulk of the farmers committing suicides and who constitute 76 per cent of farm holdings.

How Serious is the Debt Problem Among the Deceased Farmers?

According to Madhavan, *et. al.* (1998), the average debt of the sampled deceased farmers in Andhra Pradesh was only Rs. 85,242. Farmers having less than Rs. 0.50 lakh constituted about 35 per cent of the deceased sample. About 33 per cent of the deceased owed debt between Rs. 0.50 and 1.00 lakhs. Similarly, 23 per cent of the deceased had debt between Rs. 1.00 and 1.50 lakhs. Only 4 per cent of the deceased had debt of more than Rs. 2.00 lakhs. These data are suggesting that small farmers, who constituted the bulk of those committing suicide, have practically no cushion to absorb risk. Any package to resolve the crisis in agriculture, therefore, will not succeed unless 'risk' is managed well through an effective crop insurance.

Purpose of Borrowing Loans by the Deceased Farmers

89 of the deceased sampled farmers in Andhra Pradesh borrowed loans for the purpose of purchasing seeds, fertilisers and pesticides. 6 per cent of the deceased borrowed for digging bore-wells. Only 3 per cent borrowed for marriage purpose (Madhavan *et.al.* 1998).These data are contradicting irresponsible comments of some that the deceased farmers borrowed heavily either to meet personal and family expenses, or to meet expenses on vices.

Source of Finance

Banks (8 %) and Co-operatives (5 %) together provided only 13 per cent of credit needs of deceased farmers. Money-lenders (37 %) friends and relatives (35 %) pesticide and other input agents including middle men (13 %) and private finance (1%) together provided 83 per cent of the debt of the deceased farmers. Field study of Chowdary *et al.* (1998) in the distress areas in the State has indicated that tenant farmers were not getting any institutional finance. The interest charged by the private agencies was high ranging from 36 and 120 per cent depending up on the kind of sources, background of the borrower and the urgency of the need.

Why the Crisis is More Severe and Intense in Andhra Pradesh

While the national policies are the same and have similar impact throughout the country, including in the states ruled by the left parties, an important question arises as to why the crisis in Andhra Pradesh agriculture has become more severe and intense, as manifested by more number of farmers' suicides, as compared to any other State in our country? Atleast 4 answers could be given in answer, all of which have been simultaneously operating.

Firstly, the special features and stage of development of the State, and various policies being pursued by the State having a bearing on agriculture, both at macro and micro level, are important. From the data presented in Table 4.1, it could be noticed that Andhra Pradesh is well behind in development compared to other states, in several aspects. It occupied 20th place in literacy, 7th in the matter of per capita revenue, 11th in per capita budget and in banking spread and 7th in the matter of per capita power consumption.

Table 4.1 : Andhra Pradesh-Level of Development*

Item	*Year*	*Unit*	*State*	*Rank*
Area	1991	%	8.4	5
Population	1991	%	7.4	5
Literacy	1991	%	44.1	20
Working Population	1991	%	45.1	3
Cultivable Land	1991-92	%	40.2	14
Irrigated Land	1991-92	%	39.4	8
Fertiliser Consumption	1994-95	kg ha^{-1}	125.4	4
Rice Productivity	1991-94	kg ha^{-1}	2721.0	4
Food Grain Productivity	1991-94	Per capita kg	1670.0	7
Power Consumption	1993-94	kWh per capita	344.0	7
Population Per Bank	1996	No	13630.0	11
Deposit-Credit Ratio	1996	%	76.4	2
Revenue Expenditure	1994-95	Rs. per capita	1168.0	10
Revenue Income	1994-95	Rs. per capita	727.0	7
Budget Expenditure	1994-95	per capita Rs.	1536.0	1
Milk Production Per Capita	1994-95	kg	58.0	14
Annual Income Per Capita	1994-95	Rs.	7798.0	8

* *Economic Review,* April 1996-97.

Agriculturally, it occupies 8th rank in area under food grains, and 7th in production and productivity. In oilseeds, it ranks 3rd in area, production and productivity. In cotton, 3rd in area, 4th in production and 6th in productivity. In rice, it occupies 4th rank. In the matter of fertiliser consumption, it ranked 2-4th depending upon seasonal conditions. It has the dubious distinction of being 1st in the matter of pesticide consumption of technical grade with 0.82 kg. per hectare compared to national average of 0.32 kg. per hectare. India consumed 13,650 tonnes of technical grade pesticides during 1992-93, which came down to 7,291 tonnes out of 74,414 tonnes of pesticides at the corresponding national level during 1997-98. Bulk of these pesticides were being used in cotton and chillies (Ministry of Agriculture, 1998). Cost of these account for nearly 50 per cent in the cost of production of these crops. All these imply that efficiency of resource use is very low in the State with simultaneous higher cost of production.

Secondly, State Government has been pursuing the new economic and liberalisaton policies more vigorously than by any other State Governments.

Thirdly, budgetary support of the State Government to agriculture and irrigation has been decreasing over the past several years as detailed in Table 4.2. Many posts of the State Agriculture Department were not being filled for years together. Even TA bills of the personnel in the Agriculture Department at the lower level were not being timely settled. Under these circumstances, and in the face of increasing costs of every item, it is not too difficult to assess the morale and the mood of the extention personnel, who are mainly charged with the main responsiblity of technology transfer in agriculture. There was physical decline in area irrigated, both under canals and tanks. It declined from 18.69 and 15.04 lakh hectares under canals and tanks in 1988-89 to 15.39 and 7.67 lakh hectares respectively in 1995-96, largely due to failure to allocate needed funds to maintain the projects and to repair the breached tanks during the floods.

Fourthly the crisis got aggravated due to frequent natural calamities during recent several successive years in the form of droughts and floods, which have a debilitating effect, as a last straw on the camel's back on the economy of the farm families, specially those from small and marginal holdings and tenant farmers who constituted the bulk of suicide farmers.

Table 4.2 : Revenue Expenditure on Agriculture and Irrigation + Flood Control

Year	*Agriculture & Allied*	*Irrigation + Flood Control*
1988-89	241(5.7)	539(12.7)
1989-90	255(5.4)	362(7.7)
1990-91	273(5.0)	419(7.6)
1991-92	398(5.1)	811(10.5)
1992-93	474(6.6)	532(7.4)
1993-94	422(5.3)	593(7.4)
1994-95	454(4.8)	728(7.7)
1995-96	464(3.6)	840(7.9)
1996-97	561(3.9)	932(6.5)
1997-98	564 (3.9) R.E.	1056 (7.2) B.E.
1998-99	684 (4.2) R.E.	1250 (7.7) B.E.

Source : Budget papers of Andhra Pradesh.

What the Modern Technologies Imply?

Modern production technologies are heavily dependent on quality inputs like adequate and timely working capital, quality seeds, location specific fertiliser of various kinds in relation to the nature and kind of soils, irrigation, right kind of quality pesticides, labour and just and fair market both during sale and purchases. All these, obviously, need heavy investment with attendant risk(s). Use of the above inputs in a technically sound and environmentally safe manner is absolutely essential for higher production on a sustainable basis. Land relations, hence, also should encourage and ensure sustainable practices to ensure continuity of soil health and environment. With increasing capitalisation of agricultural production, market centred production, and market problems occupy the 'core' of sustenance, which may not always promote sustainability in production as presently manifested in excessive dependence on highly imbalanced use of N fertilisers compared to P and K fertilisers, excessive use of pesticides following aggressive salesmanship in the back drop of weak extension and demoralised personnel, and growing high risk crop(s) not best suited to maximise income, and increasing amplitude of prices realised by the farmers acting to the disadvantage of small and debt ridden middle class farmers. With the entry of India into

World Trade Organisation (WTO), genesis and dimensions of market forces operating on a given agricultural commodity have become more complex and less amenable for control both at macro and micro level even by the State, have become all the more critical components of modern production. Chowdary *et.al.* (1988) in their study on distress conditions prevailing among the farmers during February, 1988, following number of suicides in the State observed that inadequacies building into the system of what all modern technology constituted at present, collectively contributed to the distress among the farmers. They felt that the State (broader sense), should continue its support to agriculture sector and not reduce it in any manner, in the name of liberalisation policies, to enable absorption of modem technology even by small farmers, who constitute more than 76 per cent of farm holdings. Despite adoption of 'free enterprise' capitalist policy(ies), developed countries like USA alone is spending about $ 70 billion (Rs.2,96,1 00 crores at the present exchange of Rs.42.30 per $) and claiming that it has complied with all the requirement(s) of cutting down the aggregate measure of support to its agricultural sector. It was, therefore, logical that Chowdary *et al.*, (1998) felt that opening up of the economy to global competation via WTO has only enhanced the responsibility of State (broader sense) and not diminished it in any manner, to ensure that farmers withstand competition from global market forces and not whittled and overwhelmed by them. They felt that if global market forces overpower the Indian farmer, it would end the freedom of India both economically and politically.

Chowdary *et.al.* (1998) identified the following reasons to be the main causes for the prevalence of large scale distress conditions prevailing among the farmers preceding unfortunate suicides by some of the farmers in Andhra Pradesh.

(i) Inadequacies in input access, *viz.*, credit, quality seeds, fertilisers and pesticides at affordable prices and irrigation.
(ii) Gaps in adoption of technology and technological gaps in the matter of selection of suitable crop(s) and variety more under the dominating influence of market forces rather than

based on any sound technical consideration; failure to apply various kinds of fertilisers in right proportion based on sound principles of integrated nutrient supply system to promote sustainable higher productivity, irrational and excessive pesticides in place of ideal integrated pest management principles; lack of crop rotation, and gaps in timely and effective communication of technical information to tackle production problems as they arise, in an economical manner.

(iii) Lack of effective research backup to overcome the inefficient manner of input use in order to arrest declining productivity in several crops, to timely identify emerging problems of production, and to promote sustainable and ecologically sound production technologies.

(iv) Market related problems, both as purchasers of inputs and sellers of produce. Failure to ensure remunerative prices to small and debt ridden farmers.

(v) Lack of any crop insurance worth the name, and shifting entire burden of risk management on to the farmers, who do not practically have any risk bearing capacity.

In the light of the causes identified (Chowdary *et al.* 1998), following recommendations were made to resolve the present crisis in agriculture:

1. Continuation of support of State, in its broader sence, to agriculture, and not to reduce it in any manner to enable adoption of modem production technologies even by small farmers, whose importance has only increased and not reduced due to discharge of obligations under WTO.
2. Extension of subsidy to the extent of atleast 10 per cent of Gross National Product contributed by agriculture sector.
3. Provision of debt relief due to present heavy debt burden covering both institutional and private loans.
4. Issue of Credit Cards covering both agricultural and household expenditure to simplify the procedure of granting loans by financial institutions through well laid norms, and linking

the loan recovery to sale of produce in the market, and backed by effective crop (and not loan) insurance scheme.

5. Strengthening of Seed and Pesticide Acts to ensure quality seed and pesticides at reasonable and affordable price.
6. Ensuring supply of critical components of production like micro nutrients through public controlled agency in the market.
7. Expanding and sustaining access to irrigation to even small farmers at affordable price.
8. Regulation of production and marketing through crop specific Boards like Tobacco Board to enable farmers to withstand competition from global market forces.
9. To ensure remunerative prices to small and debt ridden farmers and to ensure price parity between the prices of agricultural and industrial products to encourage investment in agriculture.
10. Reorientation and strengthening of extension to fill up gaps in extension.
11. Strengthening of research backup to promote sustainable technology and to enable timely detection of problems arising in production.
12. Promotion of conjective use of organic manures along with their alternative sources, and inorganic fertilisers based on powerful soil testing through well laid demonstrations and critical input supply.
13. Promotion of Integrated Pest Management Practices through a network of extension, training and demonstration.
14. Policy formulation by the state to solve the basic problems of tenant farmers to ensure adoption of sustainable farm practices. Implementation of tenancy reforms to save tenants from exploitation.
15. To ensure self-reliant production system and food safety to preserve national freedom and sovereignty.

REFERENCES

Chowdary, K. R. A. Prasada Rao, and M. Koteswara Rao, Distress of farmers X-rayed-A case of Andhra Pradesh, Andhra Pradesh Rythu Sangam, Hyderabad, 1998.

Government of Andhra Pradesh, Economic Survey, Report-Submitted to State Legislature, 1996-97.

Government of Andhra Pradesh, Budget Papers, 1997-98.

Madhavan, Samiksha and Ramana, Farmers' Suicides in Andhra Pradesh, Research Advisory Group, Hyderabad, 1996-97.

Ministry of Agriculture, State Agricultural Policy and Action Plan, Government of Andhra Pradesh, 1998.

5

Suicides of Cotton Farmers

Analysis of Some Causes

C. S. MURTY

Introduction

On a scale unheard of before farmers growing cotton committed suicides in Andhra Pradesh in 1997-98. Several reasons are advanced for the suicides. Four important reasons are listed here : First, the failure of the crop caused by untimely rains/unsuitable weather led to suicides. Second, the farmers resorted to the extreme step because they were alienated from family and society in the aftermath of crop failure. Third the farmers committed suicides because they faced heavy losses/debt on account of crop failure. Fourth, the farmers used pesticides indiscriminately in the absence of a proactive extension agency even as the trader, out of avarice, supplied spurious seeds and pesticides to them. All this culminated in crop failure and the consequent suicides.

The above reasons are inter-linked. So much so it is tricky to pinpoint the reason which is by far the most important in driving the farmers to suicides. Nevertheless one should explore the ways of isolating the factor(s) that caused greater agony than others because it is important, for purposes of policy, to identify such factor(s) This is done here by adopting a comparative static framework. We take two separate groups of the highly vulnerable tenant farmers who have faced severe crop damage and analyse their reaction to the damage. Tenants among cotton growers make up one group and we know a large number of them committed suicides. The tenant farmers who suffered heavy losses in paddy cultivation in 1990-91 (both Kharif and Rabi) in a village of Godavari delta constituted another group. There is no known incident of the paddy farmers committing suicides as a reaction to the losses. The two groups are similarly situated—faced crop damage, but reacted differently to the havoc. A rudimentary comparision of the two groups is attempted to understand the most compelling reason for suicides of cotton farmers.

Only a fragmentary information is readily available on the extent of damage caused to the cotton crop in 1997 (Kharif). A realistic assessment is that the yield per acre came down from about 8 quintals to about 3 quintals. That is, actual yield was about a third of the normal yield. Data from the delta village also show that productivity there, and more generally in the entire delta, in 1990-91 was about one-third of the normal one (see panel I of the Table 5.1). Productivity was 18 bags (each of 77 kg) for the agricultural year and this compares very badly with the normal output of about 55 bags.

In both the settings there was crop failure but the reaction of the farmers to the failure was totally different. While the paddy farmers of the delta took it in their stride, the cotton farmers could not bear the failure and therefore swallowed lethal pesticides. What is it that explains the differential reaction?

Was the weather-induced crop failure the reason for suicides?

It appears that even a severe crop failure may not lead to suicides if it is due to untimely rains. To substantiate the point, a comparison of the cotton region with the paddy region is made. Both regions had untimely rains and they led to crop failure of about an equal order. Had the untimely rains and the consequent crop failure been responsible for suicides they would have affected the farmers in both the settings. But the fact of the matter is that suicides occurred only in the cotton belt and not in the paddy belt. So the weather-induced crop failure is unlikely to be the reason for suicides. The farmer is conditioned enough to bear the hardships caused by natural calamities. He does not take the extreme step when crops fail because of bad monsoon/bad weather.

Table 5.1 : Yield, costs and returns for a typical tenant farmer in the paddy and cotton belts

Panel I : Yield

	Paddy Yield (1990-91)		*Cotton Yield/Kharif 1997*	
	Bags	Value (bag=Rs.300 at 1997 prices)	Qtls	Value (Qtl = Rs.2000)
Normal Yield	55	16,500	8	16,000
Actual Yield	18	5,400	3	6,000

About 66% of the potential output was lost in both the regions.

Panel II : Loss (Yield-cost) in the bad year (at 1997 prices)

	Paddy	*Cotton*
1. Actual Yield	5,400	6,000
2. Cost	6,000[3]	10,000[2]
3. Rent (50% of actual rent)	3,300[1]	1,500[1]
Loss (1-2-3)	3,900	5,500

Panel III: Net Return in a normal Year (at 1997 prices)

	Paddy	*Cotton*
1. Gross return	16,500	16,000
2. Cost (normal year)	6,000[3]	8,000[2]
3. Rent'	6,600	3,000
Net return (1-2-3)	3,900	5,000

Loss will be about the same as the net return in a normal year.

Notes

1. In the paddy area tenants were permitted to pay half the rent in the year of crop failure. It was assumed that tenants in cotton belt were also permitted to pay half the rent.
2. Cost of cultivation in 1997 cotton crop exceeded the normal cost. Normal cost is reported to be about Rs. 8,000. In 1997 it was about Rs. 10,000.
3 The crop in the paddy belt in 1990-91 (Kharif and Rabi) was lost just before it was to be harvested. By then all the required inputs had been applied on the land and it cost the farmer Rs.6,000. It was also the normal cost (at 1997 prices).

Did the farmer's alienation cause suicides?

The differential reactions of the farmers can not be attributed to their alienation either. In the delta region farmers produce for the market. The rent that the tenant has to pay to the landlord is determined independent of caste, class or kinship. In essence, agriculture is highly commercialised. In such a setting the farmer is likely to be respected on the basis of the success he records on the farm. The farmer who runs into losses/debt is likely to be looked down. He is likely to be alienated. But such a situation did not result in the farmers committing suicide when crops failed in consecutive seasons in the paddy area.

Obviously the districts where cotton is grown are also highly commercialised. When the cotton crop failed the farmer may have got largely detached or alienated from society. But this reason by itself may not have triggered suicides because crop failure in a similar situation in the delta did not lead to suicides. In other words, had alienation been the reason for suicides it would have driven farmers in both regions into suicides. But it did not happen so. Therefore suicides can not be attributed to alienation.

Did abnormal losses cause suicides ?

The loss (yield-cost) in the paddy belt to the tenant farmer would be in the region of Rs.4,000 per acre per year at 1997 prices (given that the landlord allowed the tenant to pay only about half of the agreed rent). The loss to the cotton farmer seem to touch Rs. 5,500 per acre. The loss incurred by the paddy farmers as also cotton farmers is about equal to the net return they would get in a normal year (see panels II and III of Table 5.1). Since the loss did not make the farmer kill himself in the paddy area (even though he had no non-agricultural sources of income), it would not have done so in the cotton belt. The loss can not be a cause for the suicides of cotton farmers. One may speak of debt burden (as against loss in the 1997 Kharif season alone) as a factor responsible for suicides. After all the farmer has learnt to live with debt carrying usurious interest rates since birth. If debt prior to the start of the season had not made the cotton growers to commit suicides they were unlikely to make such a large number of farmers to end their lives in the aftermath of the crop failure. It is difficult to concur with the view that crop loss of this season alone added so much to the debt burden of so many farmers that they could no longer bear it and therefore committed suicides. If there are a couple of good harvests, the farmers would be able to recover losses, if not clear debts. If only the farmers expected to get a rewarding harvest in the near future they would not have taken the extreme step of committing suicides. But they committed suicides. What made them expect that they would not get a rewarding harvest in the near future?

Crop loss due to failure of extension and avarice of traders

The farmer in the paddy area can expect a good harvest if monsoon favours. His dependence on others in reaping a rich harvest is minimal. He is familiar with crop husbandry himself and he does not have to depend much upon extension agencies. The seeds and pesticides used by him seem to be established ones. No doubt, even the cotton farmers are experienced. But their experience came almost to no avail last season. The position of the inexperienced new/young farmer is even worse. Farmers were unable to control pests even after applying pesticides 15 to 20 times. Measures to control pests by rule of thumb do not seem to work any longer. Reports that the farmers used the pesticides suggested by traders are legion. Obviously, even this approach to controlling pests came to naught. With the new varieties of cotton and with newer pests, the need for extension has never been more. For the cotton farmer the turn of events are such that he can not hope for a good harvest in the near future even if rains favour, because he is at the mercy of extension machinery and traders. The cotton farmer has come to believe, more particularly in the last season, that the cultivation of cotton crop is fraught with several difficulties, in poor extension, and spurious seeds and pesticides. He sees no way that he can overcome these difficulties himself. Neither does he have faith in the government to root out the ills in the system in the foreseeable future. It is this scheme of things which seems to have dampened the spirits of the farmers and made them commit suicides on a large scale. This understanding makes us conclude that the need of the hour is to revamp the extension machinery and to device foolproof structures to channel quality seeds and pesticides to farmers.

6

Macro Determinants of the Incidence of Suicides of Farmers in Andhra Pradesh

A Preliminary Logistic Regression Analysis

G. Ramachandrudu, Ch. Appa Rao & B. Govindu

Introduction

Andhra Pradesh agriculture has been experiencing rapid growth during 1970's, instability during 1980's and now perhaps is in the phase of decline in the 1990's. While reasons for rapid growth in the 70's are well documented, the instability factors are analysed by Agro Economic Centre, Andhra University. The emerging declining scenario in agricultural growth in India in the 90's has been clearly confessed and cautioned by the Government of India, "The pace of agricultural growth in recent years has been constrained by a number of factors including the relatively slow growth of food grains production. The annual compound growth of food grains for the past six years between 1990-91 to 1996-97 at 1.7 per cent is lower than the annual population growth of 1.9 per cent of nineties, and therefore a matter of serious concern" (Ministry of Finance, 1996-97).

The objective of the paper is to examine the recent trends in A.P. Agriculture and examine the factors that might have contributed to the recent calamity of suicides of the cotton farmers, using logistic regression procedures.

Deceleration in A.P. Agriculture

The catastrophe of A.P. Agricultural decline is the reported incidence of 336 suicides by the Cotton farmers. Economists and

Sociologists are seriously engaged in analysing the factors that affected such extreme step by the farmers. Psychologists attribute to irreparable and accumulated social and economic loss that result in deep depression as the major causes for suicide. Therefore, the farmers' suicide can be explained by the possible cumulative factors that operate at macro level thereby trickle down to the micro level.

Bhalla and Singh (1997) have presented a comprehensive state level analysis of time series data from 1962-65 to 1992-95 on Indian agriculture. They have clearly indicated that there is agricultural deceleration in A.P. in several indicators (growth rates) ranging from net sown area, yields, output and even male worker productivity. All these indicators are reported to have declined during the 1990's over 1980's.

Table 6.1 presents five selected indicators of agricultural development in A. P. and All-India during the past three decades. It can be observed from the table that there is a clear deceleration in the growth rates of some of the most crucial indicators. There is a decline in the growth rates of agriculture, net sown area, gross cropped area, value of yield and even worker productivity. Though the trends are more or less similar at the all-India level, certain growth rates like area under cultivation have been negative in Andhra Pradesh.

Table 6.1 : Indices of Agricultural Growth Rates in A.P.

Sl. No.	*Index*	*AP*			*All India*		
		1970-73 over 1962-65	*1980-83 over 1970-73*	*1992-95 over 1980-83*	*1970-73 over 1962-65*	*1980-83 over 1970-73*	*1992-95 over 1980-83*
1.	Growth rate of agricultural output	0.93	3.61	3.08	2.08	2.38	3.40
2.	Growth rate of net sown area	-0.16	-0.29	-0.26	0.19	0.14	0.07
3.	Growth rate of gross cropped area	0.04	-0.09	-0.32	0.43	0.56	0.24
4.	Growth rate of value of yield	0.89	3.70	3.41	1.64	1.80	3.15
5.	Growth rate of male worker productivity	-0.67	1.95	1.58	0.27	0.97	1.82

Source: G. S. Bhalla and G. Singh - "Recent Developments in India Agriculture—A State level analysis in *EPW,* March 29, 1997.

Table 6.2 shows per cent share of food and non-food crops in the gross cropped area in Andhra Pradesh and All-India during the last four decades. It can be observed from the table that the shift from food to non-food crops in A.P. has been a matter of concern that too by the small and marginal farmers. While the ratio of food to non-food gross cropped area was 73 : 27 during 1970's, it is now reversed to 59 : 41 In 1990's. Though there is a similar trend in All-India, the decline in food crop share has marginally declined from 78 to 72 per cent during 1960-90. The per centage share of gross cropped area under cotton increased from 2.6 to 6.2 during 1970-90 in Andhra Pradesh, while that of all-India, it decreased from 5 to 4 during the same period.

Table 6.2 : Percent Share of Food and Non-Food Crops in the Gross Cropped Area

Sl. Year		*A.P.*			*All-India*	
No.	*Food*	*Non-food*	*Cotton*	*Food*	*Non-food*	*Cotton*
1. 1960	78.01	21.99	3.17	77.85	22.15	5.36
2. 1970	73.05	26.95	2.57	77.94	22.06	4.91
3. 1980	73.09	26.91	3.65	76.63	23.37	4.78
4. 1990	58.85	41.15	6.19	72.21	27.79	4.42

Source : Same as in Table 6.1.

Table : 6.3 presents area, output and yield of cotton crop in Andhra Pradesh during 1981-82 to 1995-96. It can be observed from the table that the average yield has fallen from 312 to 258 kgs per hectare, during 1991-92 to 1995-96.

Table 6.3 : Cotton Farming in A.P.

Sl. No.	*Year*	*Area ('000 Ha.)*	*Output ('000 bales)*	*Yield (in kgs.)*
I .	1981-82	474	663	238
2.	1991-92	707	1299	312
3.	1995-96	1059	1610	258

Source : Statistical Abstract of A.P. (1992-93).

Table 6.4 presents data on State net domestic product of Andhra Pradesh by sectors during 1980-81 to 1991-92. It can be observed from the table that the per centage share of primary sector in the State net domestic product has been falling steeply in recent years. It has fallen from 47 per cent in 1980-81 to 39 per cent in 1992-93. This clearly shows commercialization of primary sector moving away from agro related activities to industrial and service activities.

Table 6.4 : State Net Domestic Product of A.P. by Industry of Origin at 1980-81 Prices

Sectors	*1980-81*	*1990-91*	*1992-93*
I. Primary	3414.10	4696.25	4541.83
	(46.6)	(40.1)	(38.6)
II. Secondary	1216.61	2168.31	2119.96
	(16.6)	(18.5)	(18.0)
III. Tertiary	7223.95	11723.39	11776.71
	(100)	(100)	(100)
Per Capita Income (Rs.)	1380	1779	1721

Source: Same as in Table 6.3

Hanumantha Rao (1998), and Gulati and Sharma (1997) have concluded that the deceleration in the food grain output during the 1990's is explained by insufficient as well as improper sequencing of reforms. Reforms have been insufficient in so far as the exports of superior cereals like wheat and rice in which the county has a comparative advantage, continue to be restricted. Improper sequencing of reforms have a bearing on supply side factors such as irrigation and fertiliser, development of which should precede the reforms process.

Methodology

Logistic regression procedures are used to analyse the macro determinants of the incidence of suicides of farmers in A.P. Logistic regression is an appropriate statistical technique because the dependent variable is dichotomous (if incidence is reported the value is one, otherwise it is zero). The logistic regression model for the incidence of suicides is

$$Ln\ [Pi/1 - Pi] = a_0 + biXi$$

where Ln (Pi/1 – Pi] is simply the conditional odds of reported suicide, Xi represents the explanatory variables used in the equation and bi is the parameter associated with the explanatory variable.

District-wise data have been used to estimate the impact of certain selected variables of agricultural growth, social development and environmental health. The data pertain to the period of 1990's. The method of estimation in the logistic model is the Maximum likelihood estimation (De Maris, 1992).

The explanatory variables selected are (1) per centage of gross cropped area in non-food grains cultivation, (2) per centage of cultivators having < 1 hectare of operational land holding, (3) use of fertiliser per hectare, (4) rural literacy rate, (5) rural crude death rate, and (6) case fatality rate of environmental related illnesses. The results show that the non-food grains cultivation (variable 1), small land holding (variable 2) and case fatality rate (variable 6) are found to be statistically significant at 1% level and all of them have positive signs. Results are presented in Table 6.5 :

Table 6.5 : Logistic Regression Results

Sl.No.	*Variable*	*Regression Coefficient*	*S.E.*
1.	Percent of non-food crops in Gross Cropped Area	+ 0.31*	0.12
2.	Percent of cultivators having operational land holding < 1 Ha.	+ 1.20*	0.15
3.	Fertiliser used per Hectare	+0.11	0.39
4.	Rural Literacy Rate	–1.21	0.21
5.	Crude Death Rate	–0.18*	0.10
6.	Case Fatality Rate	+ 0.18*	0.10

* Significant at 1% level.

Conclusions and Policy Implications

The globalisation propaganda might have aroused false hopes on farmers and made them to do risky innovations that resulted in heavy debts that led to suicides. The interesting finding that environmental based illnesses might be on the rise because of over use of pesticides and fertilisers that pollute particularly drinking water. These diseases are high in the districts that reported suicides.

The findings of the study clearly focus towards inadequate and improper agro reforms. Without building enough social and physical infrastructure in rural areas and adequate insurance for risk, it is bound to result in great strains on small farmers' families. The agricultural extension programmes that are discontinued by the Government should be revised and expanded for keeping farmers informed about changes in the economic and physical environment.

REFERENCES

Bhalla,G. S., and G. Singh, "Recent Developments in Indian Agriculture—A State Level Analysis", *Economic and Political Weekly,* March 29-April 14, Vol. XXXII, No. 3, 1997.

De Maris Alfred, *Logistic Modelling—Practical Applications*, Sage Publications, (U.K.), 1992.

Govt. of India, *Economic Survey,* Ministry of Finance, GOI, New Delhi, 1996-97.

Hanumantha Rao, C.H., "Agricultural Growth, Sustainability and Poverty Alleviation—Recent Trends and Major Issues of Reform", *Economic and Political Weekly,* July 18-24/25-31, Vol.XXXIII Nos. 29 and 30, 1998.

Hoque, M.N. and S.H. Murdock, "Socio-Economic Development, Status of Women, Family Planning and Fertility in Rural and Urban Bangladesh", *Demography India,* Vol. 24 (2), 1995.

Parthasarathy, G., and Shameem, "Suicides of Cotton Farmers in Andhra Pradesh—An Exploratory Study, *Economic and Political Weekly,* March 28. Vol.XXXIII, No.13, 1998.

7

Globalisation of Agricultural Trade and Cotton Farmers' Suicides

P. LEELA

Introduction

The recent spate of suicides by cotton farmers focused attention of academics and administrators alike on the causes and policy measures to tackle the immensely human problem in the State of Andhra Pradesh. In this context, the present paper focuses attention on the liberalisation and globalisation aspects of cotton trade and their implications for the cotton farmers.

Area, Production and Yield of Cotton

Increase in cotton production from 5.55 million bales in 1960-61 to 14.25 million bales in 1996-97 has been mostly due to increase in yield per hectare from 125 kg/hectare in 1960-61 to 264 kg/hectare in 1996-97, as can be seen from Table 7.1. Area under cotton in India is the largest in the World but the levels of productivity are extremely low compared to 571 kg/hectare of Pakistan, 800 kg/hectare of China and 1500 kg/hectare of Israel (Bhide, 1998) forming the basis for cotton farmers' woes. Further, with nearly two-thirds of the area under cotton lacking irrigation, cotton cultivation in India is still heavily dependent on erratic monsoons.

Cotton Trade of India

Raw cotton exports are one of the principal agricultural products exported by India, while cotton imports of some varieties are necessitated by domestic consumption outpacing domestic cotton

Table 7.1 : Cotton (Lint) Area, Production and Yield

Year	*Area Million Hectares*	*Production Million Bales (170 kg/each)*	*Yield (kg/hectare)*
1960-61	7.61	5.55	125
1970-71	7.61	4.76	106
1980-91	7.82	7.01	152
1990-91	7.44	9.84	125
1991-92	7.66	9.71	215
1992-93	7.54	11.40	257
1993-94	7.32	10.74	249
1994-95	7.87	11.89	257
1995-96	9.06	13.09	246
1996-97	9.17	14.25	264

Source : Report on Currency and Finance 1996-97

output. Raw cotton exports in terms of quantity increased from 32.6 thousand tonnes in 1960-61 to 277.6 thousand tonnes in 1996-97, with a peak level of 374.3 thousand tonnes in 1990-91. Cotton exports fetched Rs. 12 crores to the economy in 1960-61 and reached a peak level of Rs. 1607 crores in 1996-97, as shown in Table 7.2. Indonesia, Japan, U.K., Thailand, Belgium and Hongkong together accounted for 81.39 per cent of the raw cotton exports of India in 1994-95 (Table 7.3).

Table 7.2 : Cotton Exports of India

Year	*Raw Cotton Exports*		*Total Agri-cultural Exports*	*Share of Cotton in Agricul-tural Exports*
	Quantity 000'Tonnes	*Value Rs.Crores*	*Rs.Crores*	*(Percentage)*
1960-61	32.6	12	284	4.20
1970-71	32.1	14	487	2.90
1980-81	131.6	165	2057	81.00
1990-91	374.3	846	6317	13.40
1991-92	106.2	305	8228	3.70
1992-93	63.7	182	9457	1.90
1993-94	297.3	654	13021	5.00
1994-95	70.7	140	13712	1.00
1995-96	33.3	204	21138	0.96
1996-97	277.6	1607	23988	6.70

Source : 1. G.O.I., Economic Survey 1996-97.
2. C.M.I.E., Foreign Trade, November 1997.

Table 7.3 : Major Markets for Raw Cotton Exports of India

(Percentage share)

Country	*1990-91*	*1994-95*
Indonesia	2.14	23.03
Japan	10.56	21.70
U.K.	1.02	14.49
Thailand	5.63	11.12
Belgium	0.82	6.31
Hongkong	10.38	4.74
Italy	3.10	3.72
France	0.17	3.38
Philippines	0.06	2.80
Bangladesh	0.54	2.12
Germany	2.29	2.04
Taiwan	4.61	1.33
Others	58.68	3.22
Total	100.00	100.00

Source : Compiled from C.M.I.E., Foreign Trade Statistics of India, 1996.

Textile exports, which increased from Rs. 73 crores in 1960-61 to Rs. 24149 crores in 1995-96, accounted for 22.7 per cent of the total exports of India. India's textile exports include exports of cotton yarn, which accounted for 8.2 per cent and readymade garments accounted for 11.6 per cent of the total exports of India in 1995-96, as shown in Table 7.4.

Table 7.4 : Textile Exports of India

(Rs. Crores)

Year	*Textile Fabrics & Mfres*	*Cotton Yarn*	*Readymade Garments*	*Total Exports of India*
1960-61	73	65	1	642
1970-71	145	142	29	1535
1980-81	933	408	550	6711
1990-91	6832	2100	4012	32553

Source : G.O.I., Economic Survey, 1996-97.

GATT Agreement for Cotton Textiles and Clothing

Two major areas of special interest to developing agrarian economies like India in the Uruguay Round of Multilateral Trade Negotiations (MTN) include: Textiles and Agriculture. Since 1961, textile trade has been subject to discriminatory quotas. The

Multifibre Arrangement (MFA) of 1974 only worsened the situation by further extending the discriminatory quota system beyond cotton textiles to include manmade fibres and cotton. As the MFA remained in force until the end of 1994, developing countries like India suffered from the discriminatory and restrictive quota regimes for a long period of 20 years. It was at the insistence of the large developing countries (which are marginal suppliers) such as India, China and Brazil among others, that textile trade came to be included in the Uruguay Round of Negotiations resulting in the replacement of MFA by the Agreement on Textiles and Clothing (ATC). The ATC is designed to dismantle progressively the MFA quotas over a period of 10 years ending on December 31, 2004. By January 1,2005 an era of duty-free and quota-free trade in textiles would be ushered between countries of the World (Bagchi, 1998). This would give a boost to the textile industry of the developing countries including India directly and to the cotton growers indirectly.

According to an estimate of the GATT Secretariat, consequent to the successful completion of the Uruguay Round of Negotiations, World trade is expected to increase by $795 billion and textile sector alone would contribute to 60 per cent of this increase (Bhorali, 1994). As India's existing and potential competitiveness lies largely in textiles, India is bound to benefit from the expansion of the horizon of India's textile export trade in the developed countries. India is a relatively low cost economy compared to South Korea and Hongkong due to its low wage costs (Kaushlk, 1995). India is expected to emerge as a major player in World textiles market, provided that India is able to seize the opportunities through a consciously developed strategic thrust.

Agricultural Trade Liberalisation

One possible reason for the suicides of the cotton farmers has been identified as the fall in the price of cotton both in absolute as well as in relative terms (Parthasarathy and Shameem, 1998). In this context, it may be noted that liberalisation of World agricultural trade tends to lead to higher agricultural prices which would benefit Indian agriculture also through price transmission elasticities for agricultural commodities including cotton. Since India is a net exporter of agricultural commodities, India stands to gain from higher World prices for agricultural exports like raw cotton.

Agricultural growth continues to remain an effective means of poverty alleviation. Liberalised agricultural trade could be used to support prices of agricultural crops like cotton and to maintain incentives for continued investment in crops like cotton. Against this background, it may reasonably be maintained that globalisation and liberalisation policies may not be the factors behind the suicides of the cotton farmers, but that the causes may lie elsewhere in the adulterated fertilisers, spurious seeds and indebtedness of the farmers.

Conclusion

From a brief analysis of the Uruguay Round of Multilateral Trade Negotiations, as well as the replacement of the Multifibre Arrangement (MFA) by the new quota-free and duty-free Agreement on Textiles and Clothing (ATC) combined with the liberalisation of World agricultural trade, the prospects for cotton in the World markets appear to be quite promising holding out an olive branch to the cotton farmers not to despair but to look forward to a bright future.

REFERENCES

Bagchi, S., "Will There Be Free Trade in Textiles?" *Economic and Political Weekly,* Vol. xxxiii, No. 27, July 4-10, p. 1684, 1998.

Bhide, S., "India's Textile Trade, Marrakesh and Beyond", *Economic and Political Weekly,* Vol xxxiii, No. 21, May 23-30, p. 1227, 1988.

Bhorali, D. and S. Sikidar, "GATT Agreement and Its Implications on Indian Export Sector", in D. Bhorali (ed) *GATT'Agreement,* Mittal Publications, New Delhi, p. 66, 1994.

Kaushik, K., *GATT Accord—The Indian Paradigm,* APH Publishing Corporation, New Delhi, 1995.

Parthasarathy, G. and Shameem, "Suicides of Cotton Farmers in Andhra Pradesh", *Economic and Political Weekly,* Vol. xxxiii, No. 13, March 26-April 3, pp.720-21, 1998.

8

Suicides of Cotton Growers

A Challenge to Extension Service in Future

N. Bhaskara Reddy

Introduction

The developments in bio-technology, tissue culture, genetic engineering will have a great impact on food and non-food crops. The new techniques are capital intensive and require knowledge and skill. Though there is a wide publicity through radio, T.V. and press, a section of the farming community is unable to adopt the latest methods due to lack of access to scientific knowledge and training. In cultivating HYVs the farmers must pay utmost care from sowing to harvesting. At present most of the food and commercial crops were covered under HYV technology. Cost of production of commercial crops is high when compared to food crops. Growing them is also risky.

Awareness of Risk and Uncertainity

The farmer is engrossed with two types of eventualities which have bearing on crop plans for the future *viz*., risk and uncertainty. In the initial stages during 1965 and 1970's the adoption of HYV technology in rice and wheat involved risk and uncertainty. But then the government's efforts through extension services and media succeeded in minimizing risk and uncertainty. The HYV technology was later extended to commercial crops like cotton, chillies, oilseeds, sugarcane and vegetables. The risk and uncertainty factors are much more associated with commercial crops. In monoculture area there is no scope for the farmer to grow a crop other than what is grown on neighbouring plots. But in the case of commercial crops, there is much scope to select competitive and remunerative crops after estimating risk and degree of uncertainty. However, selection of crop by a farmer depends upon his resource base and his technical skills. If there are eventualities in growing high cost cash crop, the big farmer will be able to cope up with losses. But if the small farmer faces this situation, he may be ruined.

Subsistence Farmer-High Cost Cash Crops

According to 1991 Census about 78 per cent of the operational holdings belong to marginal and small farmers. Past empirical studies reveal that most of these farmers have low resource base and suffer from indebtedness. They are often illiterate. When they adopt new technology, they rely on the advice of traders when it comes to use of pesticides and fertilisers. This gives rise to several complications, such as excessive use of pesticides and fertilisers. This leads to environmental degredation but also to increase in the cost of cultivation. If everything is normal the farmers irrespective of size get good yields. But if there is any eventuality like bad weather, attack of pests and diseases, there will be heavy losses. The big farmer will be able to absorb these losses to some extent. But the small fanner is unable to face these losses and falls in debt and the growing debt humiliates him. This at times makes him commit suicide.

The Suicides of Cotton Growers

Several opinions have been expressed on the causes of suicides among cotton growing farmers in newspapers and adademic journals in recent times. The causes are many like low yields, crop damage, debt, untimely rains, supply of spurious pesticides and fertilisers and finally poor extension services. Profit potential of the cotton crop lures the farmer to opt for cotton crop. Some farmers in villages like Gottipadu, sensing high risk and uncertainty conditions associated with cotton, shifted to soyabean, superior varieties of groundnut and pulses or by observing rotation of crop *viz*., blackgram with cotton, soyabean and cotton or pulses as a mixed cropping pattern (*Annadatha* August 98, p. 41).

Thereby some of the farmers in A.P. avoided losses from cotton. The individual farmer should prepare his crop plans considering his resource base.

If the farmer felt that return from cotton crop is not remunerative he might have put his land to the cultivation of other commercial crops. The data show that the area under cotton crop in A.P. increased over a period of time. In A.P. the area under cotton in the year 1970-71 was 316.2 thousand hectares and this area gradually increased and reached 655.4 hectares in the year 1990-91. The per hectare yield rate in year 1970-71 was 45 kgs and it increased to 304 kgs in the year 1980-81 and decreased to 286 kgs in the year 1990-91. In the case of other commercial crops, tobacco and chillies the area under these crops gradually decreased from the year 1970-71 to 1990-91. The compound growth rate of area under cotton for the period 1970-71 and 1983-84 was 3.1 per cent and for the period 1983-84 and 1993-94 was 4.8. In these periods the yield growth rates were 14.1 and 0.3 per cent. Thus, there is drastic decrease in yields in the period 1983-84 and 1993-94. But the yield of chillies and tobacco increased during the period 1970-71 and 1980-81 and declined in 1990-91. The compound growth rate of the yields of chillies and tobacco for the period 1970-71 and 1983-84 of these crops were 4.7 and 3.3 per cent respectively. But in the period 1983-84 and 1993-94 the per centage growth rate decreased (Table 8.1). Yet when compared to other commercial crops the farmers are choosing cotton crop because it is still remunerative. Further, even if we refer the compound growth rate relating to some important cotton growing states A.P. occupies 4th place in area and first place in yields (Table 8.2) for the period 1970-71 and 1983-84. But in the period 1983-84 and 1993-94 the State stood first in area and last in yields. A.P. farmers increased the area cultivated under cotton crop ignoring the fluctuations in yields. We also calculated compound growth rate for the period 1980-81 to 1994-95 for districts of Guntur, Warangal and Medak where unhappy things happened to cotton growers (Table 8.3). Here also there is positive compound growth rate in cotton area.

Table 8.1 : Area Yield and Growth Rates of Some Cash Crops in A.P.

Year	*Cotton*		*Chillies*		*Tobacco*	
	Area '000 hec	*Yield per hec*	*Area '000 hec*	*Yield per hec*	*Area '000 hec*	*Yield per hec*
1970-71	316.2	45	218.9	730	222.1	650
1980-81	419.2	304	162.0	920	169.4	970
1990-91	655.4	286	208.2	166	166.8	1120
Compound Growth Rate						
1970-71 & 1983-84	3.1	14.1	-1.0	4.7	-0.4	3.3
1983-84 & 1993-94	4.8	0.3	3.0	2.2	-1.7	2.0

Source : CMIE September, 1995.

Table 8.2 : Compound Annual Growth Rates of Area and Yield in Selected States

State	*1970-71 to 1983-84*		*1983-84 to 1993-94*	
	Area %	*Yield %*	*Area %*	*Yield %*
Andhra Pradesh	3.1	14.1	4.8	0.3
Gujarat	-0.7	1..4	-2.6	2.1
Haryana	5.0	0.5	4.0	2.2
Karnataka	-0.5	5.2	-4.6	7.6
Madhya Pradesh	-1.3	0.7	-1.4	3.3
Maharashtra	-0.3	1.7	-0.5	4.3
Punjab	4.3	-2.3	-0.6	7.5
Rajastan	3.9	4.1	2.1	3.5

Source : CMIE September, 1995

Table 8.3 : The Compound Growth Rates of Area of Cotton Crop in Selected Districts (1981-82 to 1994-95)

District	*Growth Rate %*
Guntur	4.2142
Medak	1.3607
Warangal	5.5428

Area under cotton has increased over the years. Along side this, the need for extension also must have increased. Extension agencies must cope with the increased demand for extension caused not only by the increase in the number of farmers growing cotton, but also by the newer and newer varieties of pests that are damaging the crop. There is no indication that the extension agency has been able to squarely face this problem of increased demand for its services. There seems to have arisen a pressure on the services of extension agencies. Under the circumstances, it is likely that the poor cultivator is sidelined by the extension agencies. Because he has a poor resource base, his access to extension services gets greatly limited and therefore he is the most vulnerable of all cultivators. When technology is fast changing, and when pests and diseases attack the farm incessantly, it is the petty cultivator who is at a great risk. Unless extension services are taken to the small farmer, he is likely to face crop failure often. Such crop failures as we have seen in the cotton belt will result in suicides.

Conclusion

The policies and technologies make the Indian farmer a gambler. New agreements like GATT new technology like bio-technology, tissue culture and Genetic engineering change the structure of Indian agriculture. But the receiver of these technologies (cultivator) does not have sufficient skill and literacy. We know that failure of cotton crop is due to weather failure and spurious pesticides and fertilisers and also due to poor extension services. The crop damages may happen to other commercial crops like tobacco, chillies, pulses, oilseeds etc. But all these crops are cultivated under HYV technology. But in the near future the internationally developed new technology will replace the existing cultivation practices in India. At this stage the Government of India should train its technical staff and farmers to face the challenges in crop production. There are two types of uncertainties, that is, nature made and man made. The nature made uncertainties in crop production should be covered by insurance. The man made uncertainty should be tackled by the efficient extension services and by strengthening existing cooperative societies.

9

Some Solutions Towards the Management of Cotton Pests

K. Aruna Lakshmi

The agro-climatic conditions of India are comparable or even superior to that of Isreal, Egypt, China and Pakistan. The area under cotton cultivation in India is the largest in the world (7.86 million ha). Yet, the mean productivity is very low in our country (324 kg/ha) and is much below the world average of 583 kg/ha (Sundaram, 1995; 1996; *The Hindu,* 1997). Isreal, a country which is smaller than Andhra Pradesh in area and with half as much rainfall produces 1686 kgs/ha. It is estimated that India will require 23-25 million bales of cotton by 2020 AD to meet the internal demand due to rising population (*The Hindu,* 1997). There is a need to enhance production and productivity in the coming decades. It is a sad irony that cotton farmers commit suicide in significant numbers and harbour fears of crop failure and losses perpetually. In as much cotton has come to be considered a make or break commercial crop, there is an imperative need for special attention by all concerned.

There are many constraints affecting cotton production. I will restrict myself to production problems in cotton associated with pests. The most important production problem after uncontrolled abiotic factors like drought and flooding is a cohort of insect pests. These pests regularly reduce the yield of cotton to levels at which harvesting is not economical.

The cotton boll worm *(Helicoverpa armigera)* has become the number one constraint of cotton production and is the reason behind the suicides of cotton growers in 1986 and 1997. It is the most devastating pest in the world. Being polyphagous it attacks a wide range of crops. Its geographic distribution is wide as the adult moths migrate over great distances. The reproductive rate of this insect is as high as 1000 per generation (ICRISAT, 1993). Tobacco catterpillar (*Spodoptera litura*) and white fly (*Bemisia tabaci*) are a few other dreaded pests of cotton which are defying many chemical pesticides. An attempt has been made in this paper to present appropriate and effective control strategies for management of cotton pests.

Pesticide Peril

Cotton occupies only 5 per cent of the cropped area but accounts for 55 per cent of the pesticides. Almost 40 per cent of the cost of cotton cultivation is due to pesticides. The easiest way to fight pests is by poisioning them. Man with all his ingenuity has waged a war on insects for the past 40 years attacking them with an arsenal of most lethal chemicals and pesticides and has failed so far to eliminate a single species.

Risk Aversion and Over Estimation of Pest Damage

In an effort to increase sales the pesticides companies embark on massive advertising campaigns. Farmers who are made aware of the pests by the chemical companies tend to over react to slight symptoms. A recent survey has shown that 60 per cent of the farmers spray insecticides even if they have not seen insect pests in their field (Escalada & Heong, 1993). They said this is to protect their crop from unseen pests. This belief has resulted in the use of highly toxic insecticides, such as methyl parathion and monocrotophos, at the early crop stages. These chemicals also kill the natural enemies and so they induce resurgence of the pest. Resurgence is the first warning and the pest should be managed at this stage as it would lead to resistant problems. Indiscriminate use of pesticides upsets the balance of life leading to large-scale outbreaks of pests. Many cotton pests are now out of control inspite of liberal application of insecticides. Cotton boll worm has become resistant to several insecticide classes. The level of resistance to chemicals is directly proportional to the amount of insecticides used. Farmers find no solution to this problem except spraying more pesticides-the insecticides thread mill. As risk aversion has become almost a habit in many farmers, it is necessary to devise techniques and policies to wean farmers away from this behaviour pattern. Since insecticides are not the answer to fight against pod borer or other pests, it is imperative to develop effective Integrated Pest Management Strategies (IPM).

Integrated Pest Management (IPM)

IPM incorporates four basic strategies for pest control *viz.*, 1. Cultural practices, 2. Biological control, 3. Use of resistant hose varieties, 4. Rational use of pesticides.

Soil Health and Pest Outbreak (Cultural Practices)

Pest outbreak is an indication that something has been going wrong. What is going wrong is always found to be the excessive use of chemicals. A famous South American ecologist, Lutzenberger (1996) says that pest attack occurs on the crops grown in unbalanced soil. Similar to what Prof G. Parthasarathy said (National Seminar on Suicides among cotton farmers in Guntur and Warangal districts) while "suicides indicate a sick society", pest outbreak indicates a sick soil. Many ecologist have established a correlation between heavy pest attack and soil types. Pests can only thrive on plants (growing in sick soils) with metabolic imbalances. Plants growing in healthy soil are well nourished and can produce their own defences against pests by producing substances antagonistic to the pests, in other words healthy plants make their own pesticides, and defend themselves. It is clear therefore, that pest attack has to do with the matabolic state of the plant. Susceptibility to pests therefore depends primarly on plant nutrition. In human terms, it has long been established that the health of the mother is the most important determinant in the health of the off spring. In plant life too, a healthy soil is essential for raising healthy crop. It is said that "pests starve on healthy plants". Therefore, plant health must be optimised for pest attack to be minimised. By maintaining soil in a good condition it is possible to have pest free crop even though it is surrounded by fields of the same crop infested with pests (Lutzenberger, 1996).

By the way we nurture our cotton crops in Andhra Pradesh, it appears almost impossible for the plants to reach nutritional balance. Farmers in their anxiety to increase yields dump in inorganic chemical fertilisers into the soil. An excess of any element causes damage. Excess nitrogen for example leads to dilation of cells, diminution of flowering and weakened resistance against diseases. Thus, excessive use of nitrogenous fertilisers is both a waste of resource and an invitation to injury. The injury is reflected in (i) lesser yield, (ii) loss of quality, and (iii) heavy pest incidence (Ghosh, 1996).

Enriching the soil with organic manure seems to be a rational approach to maintain plant health and reduce pest attack. There are many farmers who would like to use chemical fertilisers on the grounds that "there is not enough organic material to produce the manure. The real issue is not whether there is enough organic material to produce the needed quantity or quality of organic manure. The main problem is that we want to avoid the efforts necessary to collect the wastes and prefer the cosiness of a push-button system which factory production provides" (Gosh, 1996).

Crop Rotation by Soil Conditioners

Cotton is a crop which draws a lot of nutrients from the soil. The soil becomes highly depleted of many major and minor elements every time cotton crop is grown. It is therefore, important to nurse the soil back to health before growing cotton in the next season. The pulses or legumes like soyabean are ideal crops which significantly replenish the depleted nutrients in the soil. Cotton has to be rotated with legume crops in order to maintain soil fertility and subsequently plant health. The best strategy to manage pests therefore is "not looking for weapons to kill them (pests) but for ways to strengthen our crops by maintaining soil health".

Biological Control

The broad spectrum pesticides knock down the pests and their natural enemies alike. Unfortunately, the natural enemies take longer time to build up their population while the pest comes back into the picture immediately and more vigorously (in the absence of natural enemies) devastating the crop. The principle behind biological control is to introduce natural enemies into the agricultural system where a pest outbreak occurs. The enemy interacts with the pest and regulates the pest population, thereby reducing crop losses. Organisms that can be manipulated by man for pest control purposes are called biocontrol agents or biopesticides. A biocontrol agent can be a pathogen, a parasite, or a predator. A few biopesticides that are used presently for the control of cotton pests are shown in Table 9.1 Biocontrol agents are cheap to make, easy to use, specific to the target pest and are environmentally compatible posing no risk to human health.

Table 9.1 : Biocontrol Agents Effective Against Pests of Cotton

Natural enemy	*Host insect Pest*
Parasites	
Trichogramma chilonis	Cotton boll worm (*Helicoverpa*)
Telenomus remus	Tobacco caterpillar (*Spodoptera litura*)
Predators	
Chrysoperla carnea	Cotton boll worm (Helicoverpa
Ladybird beetle	Tobacco caterpillar (*Spodoptera litura*)
Pathogens	
Nuclear polyhedrosis virus	Tobacco caterpillar, Cotton boll worm
Beauveria bassiana	Tobacco caterpillar, Cotton boll worm, Cotton leaf folder
Nomuraea raileyi	Tobacco caterpillar, Cotton boll worm

The success of biocontrol agents in controlling pests may vary from 100 per cent to 0 per cent. This inconsistency is probably the biggest single factor contributing to continuing use of chemical pesticides. However, it should be remembered that a 4-way relationship exists between soil, plant, pest, and its enemy. For example, plants growing in a balanced soil are healthy, pests feeding on these healthy plants are weak and are vulnerable to diseases and their natural enemies. Biopestictides can make an impact only when the entire farming system is manipulated to reinforce biocontrol. This implies that farmers must create condusive conditions for the enhanced activity of the biocontrol agents.

India is the first nation to declare an official Policy on Integrated Pest Management (1985), however, the translation of the policy into an effective implementation strategy seems to have gone away. India's IPM programmes seem to have less impact on the farmers than those of Philippines and Indonesia. The reasons for their success lies in the effective dispersal of IPM information and practices through sustained campaigns and training programmes. For IPM to be successful in India it is necessary to adopt a multiplicity of methodologies that include (a) strategic extension campaigns like in Phillippines and Malaysia, (b) farmer field schools, and (c) farmer participatory research.

Farmer field schools and farmer participatory research programmes can encourage farmers to carry out experiments and learning about IPM. Such training enable farmers to make more rational decisions in the use of pesticides and fertilisers. In India, the Krishi Vignan Kendras are somewhat modelled along these lines. However, they have not acquired the wherewithal for effective skill and knowledge dispersal like in Indonesia. Learning-by-doing strategies appeal to farmers in developing countries.

Pre-emption: A Proactive Policy

In Indonesia, withdrawal of pesticide sales proved successful in suppressing brown plant hopper on an occassion. Though it would not be easy, such selective seasonal pesticide withdrawal could be an effective strategy to combat pests. This would require special legislation.

Region specific and crop specific mass participatory campaigns can be planned to weed out eggs and larvae of the cotton pests as a pre-emptive strategy. In 1989 in a flareup of stem-borer infestation in Indonesia, 3 lakh people (scientists, doctors, lawers, teachers and many citizens) were involved in picking stem borer egg mases from the plants. Such a programme can be planned in the case of cotton pests in A.P. as part of the now popular *'Janmabhoomi'* Programme.

Private Sector and IPM in Cotton

In Australia, pest management in cotton crop is supervised by consultants. To a question that Prof. Geoff Norton, (Director of the Cooperative Research Center for Tropical Pest Management, Australia), asked at the Ciba Foundation Symposium at Madras (1993), relating to the extent to which such private consultants can be involved in cotton pest management in India and their potential, Prof. Zadoks had an interesting answer. As Prof Zadoks (1993) put it .

> '...... *A country like India could promote this type of development. There are so many graduates of agricultural colleges who need employment and women who want to re-enter agriculture after child-bearing. One advantage of this system is that if a private consultant does the wrong thing, the farmer will not go again to him or her, and so he or she is out of business. This is not the case for governmental consultants, who may be blamed. but will be paid. Private consultants are not blamed but are no longer paid, and that is far more effective.*

This view point found many takers. It merits serious consideration by policy makers. Prof. Nagarajan of the IARI also suggests that success in IPM and biological control would come about when problem solving is decentralised. He also suggests floating of IPM enterprises by small enterpreneurs knowledgeable in plant protection practices. It could open avenues for untapped rural employment potential of IPM.

The Israeli High-technology Farming

Israel tops the world with regard to per hectare yield of cotton (1696 kgs). The reason behind this success is the use of high, appropriate and innovative technologies. The Israeli techniques of soil and water management are one of the best in the world. The Israeli scientists say "knowing exactly when to irrigate crops" is very essential in order to obtain high yields of premium quality. For this purpose they use expensive equipment like "pressure chambers", and "infrared thermometers" which readily detect crop stress (water potential of cotton is 1.6 mpa). The pressure chambers are monitored by the computers which in turn open the drip irrigation system and irrigate the crop. The total cost of this irrigation package (pressure chambers. computers, and drips) is 10 lakh rupees/ha. It is impossible to adopt such an expensive technology in India. However, the "Kuppam project" model in Chittor where Israeli collaboration has been sought through the participation of farmers and the government can be experimented in regions like Guntur and Warangal. Though capital intensive the yields compensate for the investments through huge profits and productivity. "Regional farmers crop co-operative groups" have to be formed on the lines of Israeli Kibbutz to adopt such hi-tech farming processes, with the help of governmental inputs on an experimental scale. Hi-tech irrigation packages coupled with appropriate soil and pest management strategiles are bound to increase the productivity of cotton in India.

REFERENCES

Escalada, M.M. & Heong, K.L., "Communication and implementation of change in crop protection" in : 1993 *Crop Protection and Sustainable Agriculture.* John Wiley and Sons, Chickester, New York. pp. 191-201.

Ghosh, S., "There's no salvation except through organic farming!" *The Organic Farming Source Book.* The Other India Press, Goa, India, pp. 130-137, 1996.

Lutzenberger, J.A., "Pests: Are Poisons the Answer?" *The Organic Farming Source book.* The Other India Press, Goa, India, pp. 102-115, 1996.

Norton, G., "Food crops versus cash crops" in *Crop Protection and Sustainable Agriculture.* John Wiley and Sons. Chickester, New York, p. 95, 1993.

Report, ICRISAT "Will the pod borer become the farmer's pall bearer?" special feature, ICRISAT, 1993.

Sundaram, G.K., "Fluctuating fortunes of textile industry". *Kisan World.* Dec. 1995, pp.39.

Sundaram, G.K., "Fluctuating fortunes of textile industry". *Kisan World.* Jan. 1996, pp. 19-20, 1996.

The Hindu, Survey of Indian Agriculture, "Cotton : New strategy gains urgency", pp. 107-111.

Zadoks, J.C., "Food crops *versus* cash crops" in *Crop protection and sustainable Agriculture.* John Wiley and Sons, Chickester, New York, 1993, p. 95.

10

Quest for Solutions to Pest Problem on Cotton

Use of Biological Control and Biotechnological Strategies

V. PADMAJA & A. ARUNDHATI

India has the largest acreage of 7.9 million hectares among the cotton growing countries in the world, but the yields in general and productivity in particular are the lowest. Production of cotton is a highly complex biological, physical, economical and political system in which pests play a significant role. The crop is being attacked by a variety of insect pests from seedling to final picking stages, of which six to eight are economical pests causing 50-60 per cent of the crop losses. By virtue of the evolutionary strategies, the existing pests are empowered with mechanisms for combating the man-made impediments. Hence crop protection has become a necessary agricultural practice of utmost importance for sustainable output.

Aggravation of Pest Problem

Insect pest problems have escalated with an increasing cropping intensity and with the use of agrochemicals inherent in the Green revolution technologies of modem agriculture. It has been rightly pointed out by the renowned ecologist, Odum that 'modern agriculture is based on maximum outputs in short-term with inadequate concern over long-term policies for inputs'. The consumption of pesticides for cotton is 2,472 metric tonnes per year, costing about 247 crores, which is about 44 per cent of the total pesticide bill. Pesticides perform well to some extent and that is the main reason that they are widely used. However, their injudicious use leading to the pesticide treadmill aggravates pest problem, instead of solving the same for the following reasons.

Excessive use of pesticides result in the development of resistance in the insect pests and proliferation of superpests.

Action of most of the insecticides is non-specific and kills the non target organisms in the form of natural enemies of the insect pests and the pollinators.

Hitherto minor pests become major pests because of the elimination of their predators by the indiscriminate use of pesticides.

The broad spectrum of action of chemicals result in depletion of beneficial bioflora from the agro ecosystem and favour invasion of undersirable pests.

Concept of Pest Management

Insects being highly adaptable and successful organisms are able to develop resistance not only to insecticides but also to genetically altered types. The need therefore is to keep one step ahead of insects by diversifying the available strategies. Total eradication of pests by any single means is unrealistic. Hence the aim should be not 'pest control' but 'pest management', below economic threshhold limits. The modem concept of Integrated Pest Management (IPM) is based on ecological principles for sustainable crop production and involve the integration and synthesis of different control tactics into a pest management system.

The Biocontrol Strategy

In the present context of environmental protection, conservation of agro ecosystem and stable pest management warrants an alternate strategy aiming at replenishing the badly mismanaged agriclimates with the pests natural enemies which are ecologically least desruptive and are safe to man and animals. India being a tropical country has a rich diversity of natural enemies of the insect pests. Biological pest control involve reduction in the density of pests as a result of interaction between the pest and it's natural enemy in the form of either predators, parasites or pathogens. The objective of the programme is to introduce billions of benificial micro organisms into the soil-plant ecosystem and make it difficult for disease causing organisms to grow. Biocontrol is harmless and non-toxic to many forms of life, has high degree of specificity, compatible with other integrated pest management practices and economical.

Use of the Pest Killer Fungus as a Biocontrol Agent

A fungus identified as *Beauveria bassiana* persisting in the cotton fields is sporadically growing on the notorious Lepidopteran pests of cotton, namely the bollworm *(Helicoverpa armigera),* the tobacco cutworm *(Spodoptera litura)* and the White fly *(Bamisia tabacci)*. This entomopathogenic fungus cause white muscardine disease and kills the cotton pests. Our survey in the infected fields of Khammam, Warangal and Karimnagar districts during 10th December, 1997 and 10th January, 1998 revealed the prevalence of this fungal disease on the above mentioned cotton pests (Padmaja and Ramesh, 1998). However, the natural epizootics of the fungus is not a regular phenomenon, because of it's dependence on the prevailing environmental conditions, for building up of sufficient inoculum. So, multiplication of the fungus and releasing the same into the fields help for inducting the eplizootics. The strategy is simply to enhance the natural biocontrol mechnism by supplementing with adequate number of infective propagules of the biocontrol agent. This would be the simplest ecologically and economically sound pest control device and hence sould be given more emphasis in future.

Biotechnological Approach

Development of biotechnological methods of pest management facilitate transfer of the desirable insect resistant gene(s) not only from related plants, but also from animals and microbes.

Transgenic Cotton

A pest killer gene from the bacterium, *Bacillus thuringensis* was found to be responsible for producing a toxin which is lethal to the major pests on cotton. Bio-pesticide prepared from this bacteria is in use in U.S.A., Australia and India. In view of it's sensitivity to U.V. rays, field persistence of this product is discouraging and hence biotechnological improvement was achieved by transferring this toxin producing gene to a relatively resistant bacteria.

Further, the Bt endotoxin gene of the bacteria has been successfully incorporated into the cotton plant so that the transgenic plant now possess the inbuilt mechanism for producing toxin of its own for resisting attack by the pests. However, for the transgenic cotton developed in U.S.A. to be a reality under Indian conditions needs further standardisation of techniques. The transgenic strategy appears simple but the success of the programme depends on effective regulation of the time and place of expression of the Bt gene during the developmental stages of the plant. As the product produced for insect resistance is of biological origin the question of persistence in an undegradable form can be ruled out.

On Going Work on Protease Inhibitors

In view of parallel evolutionary mechanisms in plants as well as the pests, the pest control approaches must be dynamic and exploit alternative sources. The protease inhibitors with insecticidal property were thoroughly characterised in cow pea. At Andhra University, Department of Biochemistry and Botany, the insecticideal properties of protease inhibitors from Jack fruit has been established. Further, these inhibitors have been found to be exhibiting specific affinity towards the digestive enzymes of serious pests on cotton like *Helicoverpa* and *Spodoptera.* We are looking forward to isolate the Jack fruit protease inhibitor gene for making a transgenic cotton for conferring insect resistance.

Vandana Shiva and Jafri (1998) pointed out the possibility

of Bt transgenic cotton becoming resistant to Bt toxin. Use of tissue specific promoters affecting only target pests may offer a solution to this. The suggestion of Kumar and Sharma (1994) to produce transgenics by combining two genes that kill insect pests by different mechanisms might over come the problem of probable breakdown of insect resistance. However, the opinion of Shell (1997) that the effectiveness of transgenic cotton released in U.S.A. is promising with 90-95 per cent expected resistance to insects.

REFERENCES

Jeff Schell, *Current opinion in Biotechnology,* 8; pp. 235-236, 1997.

Kumar, R.A. and Sharma, R.P., *J. Plant, Biochemistry and Biotechnology* 3.3-8, 1994.

Padmaja V. and Ramesh K., "Biological control of pests on cotton", *The Indian Express—Science Express,* 29th September, 1998.

Vandana Shiva and Afsar Jafri, *Seeds of Suicides—The ecological and human costs of globalisation of agriculture,* Published by Research Foundation for Science, Technology and Ecology. 1998.

11

Possible Reasons for the Financial Crisis of Cotton Farmers

STATE BANK OF INDIA

1. Scales of finance for cotton are adequate and are being reviewed every year as per local conditions.
2. The quantum of loan given is based on scales of finance and is considered adequate.
3. Many farmers, with an idea of making good profits are taking additional land on lease from others in addition to their own for cotton cultivation. As pucca lease deeds for such additional lands will not be available with farmer borrowers, the banks are unable to extend crop loans for such additional area/and crop loans are restricted for the lands only owned by the farmers.
4. As per our observations, the farmers are not following cultural practices like crop rotation etc., and are also resorting to rationing of crops with a view to get continuous sizeable income. This is leading to pest build up without break in the life cycle of the pests.
5. When incidence of pest is observed, farmers resort to indiscriminate and more than required/stipulated application of pesticides leading to avoidable extra expenditure as also possible resistance build up in the pests.
6. Farmers are heavily depending only on costly chemical pesticides and are not following the guidelines for Integrated Pest Management (IPM) as recommended by the Department of Agriculture.

7. Farmers are resorting to excessive and imbalanced application of fertilisers thus increasing the cost of cultivation.
8. This additional unwarranted cost of inputs like fertilisers and pesticides is not met by the banks as the costs are in excess of the scales of finance. To meet these costs the farmers are resorting to outside borrowings particularly from dealers of fertilisers and pesticides which in turn increased the supply of spurious/adulterated fertilisers and pesticides. This was evidenced in number of cases in various districts.
9. As per the RBI norms, collateral security is required to be obtained by the Banks whenever the crop loan limit exceeds Rs.26,000/-. Farmers in the absence of title deeds to their lands are reluctant to create registered mortgage over their lands as it is an expensive and cumbersome proposition with many hassles and delays involving stamp duty and registration charges. In many cases, the farmers are showing less land under cultivation of cotton to remain within the ceiling of Rs.25,000/-. This leads to the farmers not getting adequate finance resulting in outside borrowing.

PART - II

PROCEEDINGS OF THE SEMINAR

12

Inaugural Session

R. Radhakrishna

Prof. R.M. Mohan Rao

Respected Dr. Hanumantha Rao, the Chief Guest of today's function, Dr. R. Radha Krishna, our beloved Vice-Chancellor, Prof. B. Sarveswara Rao, Prof. G. Parthasarathy, the architects of School of Economics, Dr. Yelamanchill Shivaji the crusader of farmers, special invitees, delegates to the conference, collegues from School of Economics, friends from the Media, Ladies and Gentlemen. I am extremely happy to invite you all on behalf of the NABARD Chair Unit and School of Economics, Andhra University and on my own behalf to this important seminar of national significance.

I am grateful to Dr. R. Radhakrishna our dynamic, academic and highly committed Vice-Chancellor of this great University, for his intellectual, moral and financial support in organising this seminar. We are equally grateful to Prof. P. V Sarma, Director, and Members of Advisory Committee, School of Economics for their moral and material support to this seminar. This seminar is the offshoot of the NABARD's initiative in sponsoring "A Study on the Causes of Suicides Among Farmers in Andhra Pradesh". We are grateful to NABARD, for giving us this opportunity and permitting us to share the results of the study with the wider public.

The NABARD Chair Unit, set up in the Department of Cooperation and Applied Economics, by NABARD, to provide research support to its activities, is a fledgling child of five years.

This seminar is mainly intended to debate on the "Causes for Suicides Among Cotton Farmers in Andhra Pradesh". In the light of identified causes it is expected to come out with concrete suggestions to arrest this human misery in future.

The spree of suicides among cotton growers in Andhra Pradesh in general and Warangal in particular starting from December 1997 to July 1998 sensitized the academicians, public and as well as the Government to this issue. This is reflected in the form of analysing the causes as well as suggestions to avert this human crisis in future. Media has played a critical role in raising public conscience about this human misery. In this regard, the contribution of the Telugu Daily "VAARTHA" deserves credit.

An overview of studies by academics, articles and editorials in leading Vernacular and English Dailies, have thrown alternative perspectives on the causes as well as solutions to the problem. We have indicated most of these perceptions relating to the causes in our invitation letter to the Seminar and I do not want to repeat these once again, as all of you are seized with the problem. However, I have few observations to make.

At the macro level, despite achievements on many fronts during 50 years since Independence, old problems plaguing Indian agriculture have not withered away while new problems which add to the woes of the farmers came to the fore. Indian agriculture continues to be a gamble in monsoons and Indian farmers are socialised towards commercialised farming in high risk crops such as cotton through special Commodity Development Programmes under Government auspices from the late sixties and early 80s in particular. They are also used to the insentivities of the institutinal agencies of all shades in responding to their woes. Even prior to the ushering in of New Economic Policy in 1991, the farmers in general and cotton growers in particular are exposed to the vicissitudes of market economy. The 1987-88 deaths of cotton growers in Prakasam and Guntur districts are pointers to this.

The above scenario, on one hand, and continued rise in area under cotton in states, districts and villages which are mute witnesses to farmers deaths shows that farmers who are used to the ups and downs of farming for ages and commercialised farming for around two to two and half decades are unlikely to commit suicides for a single year's crop loss however devastating it may be.

A critical look at the alternative perceptions, gives an impression that what they explain is mostly causes for unprecedented crop loss but not causes for suicides of farmers. The causes must be much deeper than those identified and need contemplation by one and all.

Lack of data on the occupational status of suicide victims in the country till 1995 and pre-occupation of sociologists with dowry deaths, and suicides among house wives, which are more alarming than those of farmers, and which are on the rise after the enactment of the Anti-Dowry Act of 1977, partly account for the invisibility of suicides among farmers in the country.

National Crime Records Bureau, has published details of suicide victims by professional status for the first time in 1995. As per this 10,720 farmers are there among the total recorded cases of 80,424 suicides in 1995. Karnataka alone accounted for around 1/5th of the total suicides among farmers in the country. Though Kerala with its share of 12 per cent stood next to Karnataka in terms of absolute number, the rate of suicide among farmers (*i.e.*, number per lakh of cultivators) is the highest in Kerala. Andhra Pradesh is one among the 10 out of 19 States, with higher rates of farmers' suicides than the average for 19 States.

Regression and Correlation results based on 1995 cross-section data for 19 States in the country show that the rate of suicides of farmers are more manifest in areas with favourable ratio of area to holdings among small farmers, higher rate of suicide in the general public, high percentage of deaths to economic bankruptcy, and high proportion of area under non-food crops. The results also show that the farmers suicides are more in areas with predominance of small holdings, minor irrigation, low share of bank credit to rural areas and low share of priority sector advances to agriculture. In this connection, the negative association between area under cotton and share of rural credit in total credit also deserves attention.

Content analysis of 337 media reports has brought out economic causes as the dominant factor for the suicides of farmers in Andhra Pradesh. Among the economic factors, indebtedness, crop failure, lower prices are prominent though higher rates of interest and liberalisation also figured. Among non-economic causes, harassment from creditors particularly input dealers and money-lenders emerged as a major factor, though spurious inputs, over use of pesticides and erratic power supply also figured. Next in order are institutional factors, namely lack of institutional credit, limited purchases by MARKFED and CCI. Among the natural factors, inadequate rainfall during sowing season and heavy rainfall at the time of harvest figured prominantly.

Comparative analysis of suicide cases in Warangal and Guntur, the former accounting for 1/3rd and the later for around 10 per cent of the total suicide deaths, it is hoped will help identify the factors accounting for differential rates of suicide in the two districts. District level evidence has clearly shown that Warangal had a relatively unenviable environment compared to Guntur, in natural resource endowments like, soil types, irrigation facilities, unfavourable climate, on one hand, inadequate institutional credit support and extension services and lower market prices for cotton on the other.

Field evidence at the Village and Household level also showed that level of social integration in terms of conformity to social norms and control is higher in Warangal compared to Guntur. This also explains for the relatively higher suicides in Warangal compared to Guntur.

Social disintegration the crisis of modern society according to Durkheim refers to the weakness of the ties binding the individual to the group. The fact that the ties binding the individual to the group are more stronger in Warangal compared to Guntur do not lend credence to the hunch that suicides are possibly due to social disintegration.

Macro evidence as well as field evidence of Warangal and Gurntur district show that the youger generation of small farmers are entering the high risk commercial farming, to improve their lot despite hostile environment of inadequate resource endowments of their own, inadequate institutional support for credit, inputs, technical know-how, and marketing. They are able to survive in years of nature's benevolence. Survival is becoming critical, with increased needs on one hand and inadequate returns and increasing debt burden on the other. It is not single year's crop loss but a feeling of economic bankruptcy that is driving farmers to end their lives. This is the general scenario and Andhra Pradesh is no exception.

However, certain features of Andhra Pradesh deserve attention. Andhra Pradesh is one among the states with high rate of farmers suicide (15.10), 7th highest among the 19 States, 3rd in order next to Kerala, and Gujarat with around 44 per cent of area under non-food crops. In respect of area under cotton it stood next to Maharashtra. 3rd from top among States with high percentage of deaths due to economic bankruptcy topped the list with 1023 gms. of pesticides use per Hectare, 3rd and 4th from bottom in its share of rural to total credit.

Thus, Andhra Pradesh is one of the most vulnerable States with potential for high rates of suicides among the farmers. This calls for contemplation as to how to protect the small and marginal farmers growing high risk commercial crops like cotton. Given the current trends, it is unlikely that farmers will give-up cotton cultivation. It is equally uncertain whether the extension machinery could be persuaded to go to the door steps of the farmers. Finally, whether institutional credit agencies, highly cautious of transaction costs give a helping hand to the marginal and small farmers in the changed context of financial sector reforms is a moot point. Above all, whether the Government could take up on itself the onerous task of protecting the small and the weak in the context of the market economy is another dilemma. In this context, it must be noted that East Asian countries have never disowned the responsibilities of protecting the vulnerable sections of the farm front or to intervene in the larger interest, even within the framework of the market economy, a fact often glossed over in India. I am sure with the enlightened lead by Dr. C.H. Hanumantha Rao and under the stewardship of Prof R. Radhakrishna, and other elders and experts, the Seminar will be able to come out with meaningful suggestions to the farmers, institutional agencies and the policy makers to put an end to this human misery.

May I now have the privilege to introduce Dr. C.H. Hanumantha Rao Chief Guest of this morning It would be presumptuous on my part to introduce Dr. Hanumantha Rao who is too well known to be introduced. Yet I would not like to lose this golden opportunity of paying my respects to this veteran and erudite scholar. He was former member of Planning Commission, Director, Institute of Economic Growth, New Delhi, President of Indian Economic Association, Indian Society of Agricultural Economics. He was Chairman of several Committees constituted by Government of India. He was on the Prime Minister's Advisory Council of Economists. Recently, as Chairman of the Fertiliser Committee Dr. Rao has made valuable suggestions to overhaul the Fertiliser Policy of Government of India to make it more farmer friendly.

He was an voracious reader, an outstanding scholar of great repute and above all a humane person, with unbelievable commitment and concern to improve the lot of the vulnerable sections of rural India. Sir, we are extremely grateful to you for being kind enough to grace this occasion and to inaugurate this National Seminar of great concern and significance. We welcome you sir once again.

Prof. R. Radhakrishna

Prof. Hanumantha Rao, Prof. Parthasarathy, Prof. Mohana Rao and friends, at the outset I congratulate Prof. Mohan Rao and the Members of NABARD Chair Unit for organising this Seminar on the deaths of farmers in Andhra Pradesh. The deaths of farmers has become a major policy concern in the recent past. More than 300 farmers committed suicides in 1997-98 and more than, may be 1,000 farmers were affected by the cotton crisis. Alternative explanations have been offered to explain these suicides. These explanations have been drawn from alternative perspectives. These perspectives are, market uncertainties and market failures due to commercialisation and liberalisation, failure and inadequacy of market supporting systems, particularly, absence of insurance market, failure of formal credit agencies to respond to the changes, failure of collective institutions, failure and breakdown of community systems particularly family system to overcome the crisis.

It is also viewed that the penetration of technology has led to greater yield uncertainity in the 1980s, and also in the recent period. This also could be a cause for the present cotton crisis. There is also a change in the motivation of farmers, farmers are no longer averse to take risk. Some explanations have been provided that farmers are willing to take risks to come out of the debt trap. These explanations require critical assessment by scholars. In this connection, the NABARD Study assumes prominence and I am sure it will throw insights into the deeper causes underlying the crisis.

I am particularly happy that Prof. C.H.Hanumantha Rao agreed to deliver the Inaugural address. When Prof. Mohan Rao told me that Prof. Hanumantha Rao agreed to visit Visakhapatnam and agreed to deliver the inaugural address I did not believe. Rarely Prof. Hanumantha Rao concedes to requests but I do not know how Prof. Mohana Rao could convince him and bring him here. Most of you remember, he is a very eminent social scientist, with a vision and concern for the poor. He has been influencing the National Policies since early 70s. He was advisor to the Prime Minister. As most of us know he was also member of the Administrative Reforms Committee. He was member twice of the Finance Commission and he was also member of the Planning Commission and he was the Chairman of the Rural Landless Labour Enquiry Commission. That is very important position through which he has provided insights into the functioning of the labour market. As a member of the Planning Commission, he introduced the Integrated Rural Development Programme (IRDP). He has also, tried to introduce Non-Governmental Organisations (NGO's) into the planning process. He gave support to the National Non-Governmental Organisations as a member of Planning Commission. You also know that recently he was the Chairman of the Tehri Dam Rehabilitation Programme. Whenever there is a controversy, the Prime Minister looks to him to solve the problem. Once without consulting him, the Prime Minister announced in the Parliament that Prof. Hanumantha Rao would be the Chairman of the National Water Dispute. But Prof Hanumantha Rao was preoccupied with Tehri Dam work as well as with the Fertiliser Pricing Committee work. Both were appointed by the Prime Minister and Prof. Hanumantha Rao declined the offer. That is why I was hesitant when Prof. Mohan Rao said that he would invite Prof. Hanumantha Rao. We are very grateful to you, Sir, for agreeing to be here with us.

We are also very grateful to Prof. G. Parthasarathy for agreeing to participate in the Programme. Prof. Parthasarathy is the first person, the first social scientist, to react on anything that is connected with human misery. He made some analysis of the suicides of cotton farmers and also published an article in the *Economic and Political Weekly* giving his own view points on this important issue. I now request Prof Hanumantha Rao to deliver the inaugural address.

Prof. C. H. Hanumantha Rao

Prof. Radhakrishna, Prof. Sarveswara Rao, Prof. Parthasarathy, Prof Mohan Rao and fellow participants in the Seminar, let me at the very outset congratulate the NABARD Unit of the Department of Co-operation and Applied Economics for taking this initiative to organise a Seminar on the agricultural scene, on the agricultural problems, highlighted by the suicides of farmers in Andhra Pradesh. I have seen the keynote paper here, as well as a number of papers by other scholars and I must say that I have rarely come across, in the recent period, among our professionals such a response for the challenges thrown by transforming agriculture. The papers are of a high quality, in the intellectual and technical sense and also show a great social concern. I am particularly happy because the overall approach which comes out of the Seminar papers is very balanced and practical and I am personally in agreement with the kind of approach, as you will see in the course of my own observations. I am very happy and indeed very proud that some of us in Andhra Pradesh have responded to the challenge so constructively.

While I appreciate the kind words spoken about me by Prof. Radhakrishna and Prof. Mohan Rao, I do not know how I could miss this Seminar, I could never have missed this Seminar unless I was away, for whatever reasons, from the Country or Andhra Pradesh for a long time. This Seminar is in response to events which really highlight the basic problems of Indian agriculture in the context of economic liberalisation. There are some unresolved issues in this regard and I thought this will give me an opportunity to interact with you, friends, and to clarify some of my ideas on this. Prof. Radhakrishna raised a valid point about the farmers' behaviour, particularly small and marginal farmers. Generally we think that those who have weaker wealth position have a high risk aversion and they are not prepared to take risk. That is an assumption. But all the facts show, not only from Andhra Pradesh, but from many parts of the country, that small and marginal farmers take risks in the same way as large farmers do. Now we have the whole Green Revolution history right from late 60's to this day, particularly affecting the farmers in the rice and irrigated tracts of our country, whether it is Punjab or Andhra Pradesh or Tamil Nadu, Haryana and now Uttar Pradesh and West Bengal. In all these places small and marginal farmers have been actively participating in the Green Revolution. They are a little late comers. We have seen the small farmers adopting it, they have assessed the degree of risk involved, profitability involved; and with adequate extension and credit support they have been participating in it fully. In fact, all the data show that the level of fertiliser use among small and marginal farmers for wheat and rice is no less than that among large farmers. In fact, they are slightly higher in some cases. Therefore, Green Revolution experience shows that not only small farmers have been willing to take risks in the same way as the large farmers do, but also have somehow been able to withstand the risks also. We have not come across suicide cases among farmers in the Green Revolution areas of irrigated tracts of rice, wheat etc. The fertiliser and pesticides doses are very high compared to the traditional varieties, but even then the small and marginal farmers have been able to withstand whatever risks that have been associated with the Green Revolution Technology in these irrigated pockets or high rainfall areas.

In the unirrigated or dry areas, cotton has been grown for ages by the farmers, even by small and marginal farmers. When I did my field study in Rayalaseema way back in late 50s I found small as well as marginal farmers and everybody was growing cotton, jowar and groundnut in a way. In fact,. 14 to 16 crops are grown in an year in that area, Ananthapur especially. They have their own strategies of coping with risks through multiple cropping, etc. Thus, growing of these commercial crops under unirrigated conditions has been known and practiced by even small and marginal farmers for ages. So what is new in the whole thing which in fact led to extreme actions by the small and marginal farmers in some parts of Andhra Pradesh and in other places of the country also?

As Prof. Mohan Rao mentioned farming has now become commercial. Of course, a crop like cotton has been a commercial crop for ages and people have been growing it. But, these farmers have been exposed to high risk under high degree of uncertain conditions. Risk has been there. They have been exposed to risk earlier also. Now, it is the consumption of heavy doses of inputs along with high risks which has really created this new class of issues. But, even here small and marginal farmers have been undertaking risk. This is a case where the willingness to take risk, has not been associated with the capacity to bear the risk. It has not been associated with the institutional framework of coping with the risk. But even this experience is somehow understandable on purely rational economic grounds. If you look at the data on cotton area, price and yield in Andhra Pradesh, in Warangal and Guntur, you will find that the recent increase in the area particularly the last decade under cotton is perfectly understandable in terms of the usual supply responses that all of us are familiar in Indian agricultural economics. I think, basically it is price driven. The prices of cotton in the last 12-13 years, are almost tripled. It is a 240 per cent increase, even if you allow for variation and take only the triennial average, 12 years ago, and the last three years, not the disastrous year of

1997-98, I am excluding that as far as price is concerned, the price increase has been phenomenal. In fact, yield increase has been negligible in this period. As one of the papers prepared for the Seminar has brought out very clearly that in the last decade or 12 years, the yield increase has not been that much. So basically this increase in cotton area has been price driven, rather than technology-driven or yield-driven. The price response to rising prices by the farmers and acreage response have been very well documented. The author states the reasons on the basis of his experience and facts, and also if you see the yield increase, it has also been not very significant in the last decade but there has been substantial increase in the earlier decade *i.e.*, in the 70's and 80's. But when you compare between these two periods, beginning from 1970-71, probably 70s and up to mid 80s as one period, and the last 12 years, you can say you will find a variability in cotton yields in Andhra Pradesh, yields have declined. Of course, variability is still very high when compared to paddy or other irrigated crops. The risk is greater no doubt. But the farmers also are guided, by their past experience. If you look at the standard deviations of the annual variations which is a good measure of fluctuations, in output, prices, in Warangal, the standard deviation of annual changes has declined from 135 in the earlier period to 41. This is surprising. But the data reveal that. I am not including this trecherous year of 1997-98. It is a peculiar year. There were very specific circumstances in this particular year. The farmers are guided by the earlier experience and were caught unawares. Now you have this kind of story evolving itself. But in Andhra Pradesh as a whole the variability has declined. It has been reflected in the standard deviations declining from about 48 to 28. In Guntur also it declined from 90 to 38 so the farming community as a whole, including the small and marginal farmers, are under the demonstration effect from large farmers and others. Otherwise, they would not have adopted the technology. So the last decade's experience has been there. This is not to suggest that the variability or uncertainty is not really an important issue. As I said earlier, in the absolute sense, the variability for this crop, is still much greater than for some other crops and particularly when it comes in the context of high inputs. The input level has increased in the last decade, particularly of pesticides and fertilisers. So in the absolute sense, the risk of losing large amount of money has increased over a period of time. This you have to keep in mind.

Now this tragedy that occurred last year has understandably evoked various kinds of responses. I want to see the response from professionals, from academicians, but we have had responses coming from NGO's. There have been responses or lack of response, we can say, from the Government. In fact, the total inactivity, confusion and a kind of irrelevance if you are really very serious. We are here to discuss all those aspects on Government Policies. To some of the environmentalists and others who have been otherwise, against opening up of our economy against Globalisation in that sense, have been seeing this tragedy as vindication of their stand, that we should not go in for Globalisation. That we should not even go in for new technology like Bio-technology; because no technology is perfect in the short-run. It is a continuing process. Any technology introduced gives rise to some consequences which are not foreseen. Therefore, there is a continuous research to improve upon the technology. So we have to look at the prospects from the technology considering the possibility opened up by the biological sciences.

There is a very good paper which argues for continuance of the policy of Globalisation and it brings out how cotton is going to take favourable prospects not only favourable terms of trade but also the export prospects in future. I think this is important. I may say the very same cotton which has led to suicides by the poor farmers can become, if properly managed, a source for eradication of poverty. Now I am talking of cotton not just as a crop, but in the larger terms of cotton based industries, textiles, garments etc. New prospects have been opened up after the Uruguay round of negotiations. In Bangladesh the recent experience shows that for the first time many poor, migrated in large numbers from rural to urban areas were working in the garment industry and they were listed above the poverty line because export prospects have increased for this industry very much. And as you all know, I am sure you are going to discuss policies for a developing economy like India for a long term favourable position so far as cotton textiles are concerned because of phasing out of the multi-fiber agreement. That is going to cause the crisis and improve the market prospects for our good. Therefore, Globalisation I think offers opportunities for increasing the incomes of the farmers through technological change, because we have to compete with the other countries, not only in cotton production but in cotton textiles and in garments. You have to compete with other countries and therefore you have to introduce cost reducing technologies, you have to introduce practices which are acceptable to the consumer abroad. So all these vistas are opened up by Globalisation and I do not think the right lesson from the incidents is to work against Globalisation or for Globalisation. That will be a wrong signal to offer and about technology I had already mentioned.

In fact, I had an occasion to discuss this issue about bio-technology particularly the characteristics in so far as they reduce the need for pesticides, chemicals in general. I am talking about the prospects of the technologies which have been actually offered so far. If you happen to read the paper from the world renowned Biologist Prof. James Peacock of the Australian University who was invited to present this at one of the International Conferences of Agricultural Economists and I was also asked to comment on that as an Economist. It gave me an opportunity to read that whole thing and assess the prospects that bio-technology offers and I was convinced and I have no reason to revise my understanding on the discussion and it offers enormous prospects for evolving technologies and practices which can combat pests and diseases by building those properties into the seed itself apart from other practices like Integrated Pest Management (IPM) and other things. Bio-technology offers such prospects. But what is our attention to Bio-technology? Once, a few years ago, when I invited few friends in Hyderabad and I asked a very senior scientist "What do you think of the prospects for Bio-technology in agricultural research?" He said "If I have any say I would not allow a single pie for Bio-technology". The same scientist is now in-charge of a mission on Bio-technology some where. Even the ICAR's very senior scientists, I don't want to name them, a decade ago, I kept on asking these people and they say "Bio-technology is just a slogan nothing will come, we are not prepared for that". Where is our infrastructure and what are the prospects for that kind of a thing. When a scientist looks ahead, that is what we feel generally, operates on the prospects offered by basic sciences, not technology that is available in the market. So, now, of course in the last few, one or two or three years, there has been a greater awareness and a number of departments in Bio-technology have come up. But still, our investment in agricultural research is only 0.5 per cent of the value of agricultural GDP as compared to 2 to 3 per cent of agricultural GDP being allocated by several countries. Even the ICAR, in their Ninth Five Year Plan, projected this at one per cent, which is a very modest amount of agricultural GDP, which of course is doubling. But, I do not think that is a big amount at all. Further, the manner of reducing fiscal deficit is such that the first casualties are research and extension. Some of these essential outlets become casualties in our efforts to reduce fiscal deficit. Fiscal deficit should be reduced but in what way-by mobilising more revenues or by cutting down on expenditure?. So we do not know whether we are giving adequate attention to the newly emerging technologies, to the newly emerging areas to be able to benefit from Globalisation. So far as economic reforms which have been implemented so far, are very easy kind of reforms, which can be implemented by stroke of a pen by liberalising this, by opening up this, or by licenses, or by liberalising exchange rate, or things like that. But, where effort is involved, where larger outlays are involved where re-organisation of institutions is involved, our efforts have been very slow. There is not even realisation, on the need to undertake those reforms which require essential expenditure and organisational and institutional restructuring.

The other response should be from Government. The response from the Government as I said is a lack of response. Now credit comes as one of the very important aspects. The whole farming sections, whether small or marginal or big, or all of them put together, depended on institutional credit in these areas—

Warangal and Guntur, only to the extent of 20 per cent or so, particularly, the small and marginal farmers and those who have committed suicides. Therefore, for that kind of a group, the institutional sources as I pointed out, hardly supplied 10 per cent of the total need for them and the remaining 90 per cent came from money-lenders who are also traders in cotton and pesticides. They also do extension work, which the papers reveal. Now this is a peculiar kind of situation where the State is abdicating its responsibility in the context of economic liberilisation. By State I do not mean State Government, I mean Government whether centre or state. The priority sector advances for agriculture, from commercial banks were supposed to amount to 18 per cent of that bank credit, has come down to 14 per cent 3 years ago. Now I understand it is 11 per cent or so. I happened to be on the Board of the RBI and I get all these kinds of observations and assessments. Many of these Committees including the Gupta Committee which has been discussed very widely, say there is an increase in credit, in absolute terms. There is an increase we all know. For every year there is some increase and we even do not know what is the rationale for this 18 per cent? I asked the other day what is the basis? In fact, a very good study, I suppose, is going on in the RBI to analyse the supply constraints, supply factors for credit and to come out with some recommendations. R.B. Gupta Committee does not really address this issue, of supply constraints in credit at all. Now, I understand Mr. M. Narasimham, Chairman of the Financial Sector Reforms Committee, the other day, in an informal meeting told me that when he was the Governor it was fixed as 18 per cent for agriculture and 40 per cent for the whole priority sector on an empirical basis. What is empirical basis? Empirical is a very respectable term in our research you know. In the last three years what was the actual position? The basis was just casual. Whatever was there in the last one or two years before this decision was taken became the basis, became a principle for allocating credit. Now how can you make it a principle in that case? You should assess the contribution of agriculture to National Income, *i.e.*, 30 per cent. You should also see their emerging requirements.

Capital intensified agriculture has been globalised now. Horticulture and dairying both require capital investment. So what is the basis? All these questions need to be raised and the whole issue of supply of institutional credit for agriculture has to be put on a sound footing. Now it is not merely the decision regarding the understanding these allocations. It is also the institutional set up. One reason why this ratio has declined even from 18 to 11 per cent or so, successively is a result of, one can say, Economic Liberalisation. Because all these, Banking Reforms Committee's which have been sitting and making their recommendations have been to raise the profitability of banks. There are so many ways of reducing cost, reducing non-performing assets and raising profitability. There is no doubt about it. But in the drive to raise profitability many banks have been shying away from agriculture, because in agriculture we have to deal with large number of small accounts. While getting deposits, or advancing credit a typical banker at the counter would prefer a small number of large customers where the transaction cost or the unit cost would be lower. Now this is an area where Government may come and say their transaction costs are high up to this point. May be a small account costs you 30 per cent higher or 50 per cent higher within the total amount and you will subsidise and the bank would not be inhibited in that case. Then they would not be losing on that account and interest rates can be raised. Personally, I have no objection when the market rates are 36 per cent or like that, I do not see any reason why a commercial bank credit should continue to be heavily subsidised. The subsidy can be lower. The interest rate should be lower than the market rate, there is no doubt about it. But, there can be some additional costs on account of transaction cost. So there need to be a whole revamping of our credit institutions, so that not only a larger share of institutional credit is allocated but the delivery system as such is to ensure that credit actually reaches the needy farmers. In the delivery systems there are several innovations such as group lending. Bangladesh Grameena Bank Experiment is there, a number of self-help groups have come up in our Country particularly the women groups DWACRA and other self help groups have come up. Our own experiments have improved our delivery system, to involve the borrowers themselves as groups in lending operations so that repayment is effective.

Then there is an issue of insurance. Crop insurance, which needs some discussion. I suggest it needs a whole Seminar and I think it will be rewarding, if we, can collaborate with other institutions in Hyderabad or here, for a good Seminar on the issue of crop insurance, or any other insurance for agriculture. You all know I happened to study the subject too long ago when Dr. Dharam Narayan, a very eminent Agricultural Economist, was the Chairman of the Agricultural Prices Commission in 60s and 70s. There was a committee appointed by the Government of India in 1970 on Crop Insurance. I had an occasion to be a member and study the issue at that time. The result we arrived at it is good. I myself have a copy of that I will try to get it out. A number of other studies have come up after that. Prof. Dandekar wrote a series of articles and a number of other responses are there. Insurance in its strict sense may not be possible for the highly uncertain agriculture we have. What is the insurance principle? The principle of insurance is, basically it should be self paying, self liquidity. It should not be a subsidy. We have to make a distinction. If you would not give a subsidy, do not give, but insurance is a different thing. Insurance is a paying proposition. The principle is very simple, each individual pays a small amount. It is a kind of a loss to him, a very predictable, a certain loss, but a small loss, in anticipation of a very uncertain gain, a large gain, very uncertain. But he is prepared to part with a small amount he does not mind so long as he knows that one day it may be available, it will become necessary. So in purely actuarial sense it is safe. But in agriculture it is not like that. In agriculture when crops fail, they fail in the whole region. If the harvests are good, they are good every where in that particular region at least. Yield variations are not much even if you take larger regions like Andhra Pradesh. There are good years and bad years, even for the country as a whole. There are bad years to the extent of agricultural production falling by

10-15 per cent and rising by 15-20 per cent. So crop insurance to be effective should involve a very high premium may be 10 per cent; we calculated at that time; 10-15 per cent of output we have to pay as premium so that in a bad year, the frequency of which may be one in 5 years, even if you are compensated to the extent of 50 per cent of your yield loss, your premium has to be very high. I do not know what 10 per cent means. It means, you know 10-15 times of land revenue. When people are reluctant to pay land revenue which is less than 1 per cent, of their gross output, how they will be willing to pay 10 to 15 per cent, of their output value for the sake of the insurance? Even if that is true, they may say why not! I save this amount myself in the bank and use it in a proper way. You can say that even if he does not know. If in the long run a group of farmers had to pay for themselves, like this, then why not encourage them to save in bank accounts and in other institutions so that the money is available in a bad year. So I was just raising these issues for your consideration but there are other reasons and uncertainties which are avoidable.

In this whole area of risks there are many manageable areas or elements. One of your papers, again, has mentioned comparing the losses in paddy and cotton, where the losses are similar, the small and marginal cotton farmers take extreme step of committing suicide.; Why this is so? I think the hypothesis advanced, say, that there is a large amount of uncertainty and uncertainties are of different kinds. I am talking of those which can be avoided. If extension is provided by the Government, then the harassment and manipulations of a trader-lender can be eliminated; that can mean a substantial reduction in the cost of cultivation to the farmers and also it would mean the right knowledge extended to them. They know what is the right seed, what is likely to be the right yield within certain limits. So these uncertainties are avoidable. The risk avoidance is a very big area and in any seminar that we are going to have I think we should discuss the strategies for risk avoidance. Assumption of risk is a different thing. Those risks which are not avoidable, for instance, irrigation development also avoids risks of crop yield variability over a period of time and there are so many others. Your minimum support prices, which were introduced earlier, is also a risk avoiding measure. Farmers are assured, they are given an assured price, otherwise they will meet uncertainty. So we should explore all these possibilities for minimising risk, or avoidance before we discuss those risks which are to be assured for insurance or other things, that means they are not avoidable. They have to be undertaken to be addressed to some kind of a collective support, whether crop insurance or some other insurance. In this context why only farmers, alone bear the risk. This is a peculiar case, very tragic case, where so much profitability is there and so many intermediaries operate including, particularly those private traders and those who are advancing credit and also providing extension and goading the farmers to apply more pesticides and things like that but they do not bear any risks. This is one reason why there is a big incentive to teach the farmers or to tell them all kind of irresponsible things, because they are not going to bear the risks. They know ultimately the risk will fall on the farmer himself. In fact, 36 per cent rate of interest includes the risk also, as our theoretical literature show. So he is well provided on that account. Even under traditional share cropping, the whole share cropping institution was evolved for spreading risks between the land owners and tenants. I am not arguing for share cropping in the case of cotton. No. It cannot be. But there can be other methods for risk sharing. Why not the Government take a part of these profits from the traders and the inputs suppliers who are making such huge profits, on account of those who are making cultivation of this crop. Some kind of insurance has to be provided for these farmers. Why not the Government itself allocate funds for the purpose? After all, cotton exports, as I said, globalisation that is coming up is opening very large prospects for our agriculture development and for our economic development in general. Country as a whole will be benefiting from this. Why not we earmark certain funds for risk sharing? So therefore, assuming risks should not fall on the farmers alone, the rest of the beneficiary should be involved. That is how we can spread the risk over larger number of parties than the farmers themselves.

I think I have spoken for quite a long time. I would like to thank you the organisers of this Seminar for giving me this opportunity to share my views with you and I would like to congratulate again the Department for organising the discussion.

Prof. G. Parthasarthy

After the erudite speech of Prof. Hanumantha Rao I don't think I have much to say except the fact that suicides represent a sick society. Society must be getting sicker. Otherwise life is valuable. If life is valuable suicides are not committed. There is a value which indicates a particular value system in society. The fact that people are committing suicides that they are ending their lives means society is getting sicker, it is losing its integration and solidarity. So these are the aspects to which one has to refer. Possibly this could be done by seggregation of the farmers into those who have committed suicides and those who have not committed suicides. After all the number who commit suicides is far less than the number who do not commit suicides. Go into the causes of why people commit suicides. We must consider, the family and the family circumstances also, the circumstances under which a farmer commits suicide is quite relevant. So I would like to emphasise that suicides is not a simple affair, it is an indication of sick society, and sick society is an indication that people are losing the value in life; life is valuable and that sense of being is lost by the individual. When does one feel that life is not valuable ? May be when he is not well integrated with the family, well integrated with the group. So these are the questions that sociologists have to answer. Otherwise, the critical issue which Prof. Hanumantha Rao emphasised is quite relevant. How to avoid risk and how the community could help the farmer to bear this. These are important but then the most important thing is : is society getting sicker that is the most important thing that one has to consider. These are the answers which sociologists could provide.

Prof. B. Sarveswara Rao

Chairman, Mr. Vice-Chancellor, Prof. Hanumantha Rao, Prof.

Parthasarathy, Prof. Mohan Rao and friends. I think I have a little hesitation to speak now and to entrench on the valuable time of the Seminar, discussion and the sessions. I came here mainly to listen particularly to my great friend Prof. Hanumantha Rao who is a great authority on these subjects, who can bring to bear on the discussion not only whether it is a National and the Global perspective as well as the particular perspective. I am very happy that I am present here to listen to his very illuminating address. I am quite sure Prof. Parthasarathy will in due course of the session will enlighten us much more than he did in his very brief remarks. Prof. Mohan Rao came to me about a few days back and informed me about this Seminar. In fact, he came to me about two months back when he was doing the study, I mentioned two or three things with my very limited knowledge of these matters. I asked a question, of what exactly there was in social sciences in a discussion of this kind of a subject-suicide? Is it mainly the concern of social psychologists? or is it also an area for social scientists? and in that context, I said we have a very important role because nothing will happen, no individual event will happen without a social context, without an environmental context. All individual events also abnormal events of this kind have to be examined within the framework and context of a society or a global situation. So from that point I think we have a very important role to play as social scientists. So far as economists are concerned, I mentioned that our particular focus should be on risk and uncertainty analysis. In modern commercialised agriculture and I am happy that Prof. Hanumantha Rao touched on this issue of risk and uncertainty analysis. One may find very curiously that in terms of modern farming whether it is cotton or some other crop the risks are the same. You cannot make much distinction. It is not therefore the risk and uncertainity analysis that may explain ultimately the situation. The sociologists, I think the social anthropologists have to come forward and analyse the institutional structures, the caste structures, the power structure in society. It is here I think that sociologists have a very vital role to play. Finally, when it comes to the question of social psychology and culture, cultural dimension has to be examined. I think it is equally important whatever be the provocation, economic provocation, distress and agony and other things. Why should a man finally decide to end his life? It is a very serious question. It may not be directly related to either economics or sociology it has something to do with culture and psychology. I am very happy, I mentioned actually a good number of individual case studies might throw more light on these problems than an analysis of the situation at the sectoral level or the regional level or the global level. That is one remark which I wanted to make.

A second remark which I wanted to make is, apart from analysis by distinguished sociologists, a medical scientists may make a distinction between natural deaths and unnatural deaths and in the analysis of unnatural deaths, the medical scientist have an interesting group of deaths. One interesting group is that death is not preceded by a long, short period or a long period of depression. Depression is a technical and scientific term. It is a medical term. A depression usually precedes death, but in the case of some of the death's, as in the case of cotton farmers, perhaps a short or long period depression did not precede. So what happens was it is what is called an impulsive death. The impulse itself arose out of a very peculiar situation that the farmers not only immediately incurred a serious loss, economic loss and distress, but I think they felt that a total lack of confidence in the society and the Government to come to their help. It appears to me that it is a case of loss of confidence, in either the Government or in the society which made them feel that next year future is bleak. There is a no chance for a retrieval, there is no chance of a recovery of the lost ground. That kind of a psychological situation made them go in immediately for suicides. Suicides are not after all very uncommon. They are happening in most sectors. As the statistics show 80,000 people die of suicide every year out of which 10,000 are farmers. It is a different thing altogether. But why should a man go to the extreme of ending his life. It looks as though the farmer felt total loss of confidence. Will the Society come to my help this year? Next Year? Will the Government come to my help? This year? New year? Some psychological breakdown has taken place and therefore this kind of a thing has occurred.

I came here, to listen and to learn because I think it is a most fascinating subject for social scientists and an inter disciplinary perspective is absolutely essential for an understanding of a subject like this and I congratulate Prof. Mohan Rao for producing one of the best analytical papers which I could think of. I read it carefully and it is so fascinating to me. His analysis and style of presentation both of them I liked immensely. I would like to publicly congratulate Prof. Mohan Rao for this venture. I am very happy that the other papers which are circulated also are very illuminating papers. I am very happy because this University never ignored the social concern. Prof. Hanumantha Rao made a reference to that. Andhra University Department of Economics to begin with, in fact, even 40 years ago, when we started the Department of Rural Development and Co-operation and Applied Economics obviously, it was our social concern. Otherwise we would not have started the Department of Rural Development and Co-operation and even right from the beginning I think our intellectual studies have always been charged with social concern. This I am very happy that Hanumantha Rao has noted and made a reference to this in his very illuminating address and I do not think I have anything more to say I am very happy and grateful to Prof. Mohan Rao and his colleagues for giving me an opportunity to be present here for so long as I can stay here depending on my health and learn something.

Prof. P. V. Sarma

Respected Vice-Chancellor, Prof R.Radhakrishna beloved Prof. Hanumantha Rao, Prof. R.M.Mohan Rao, the Seminar Director, senior Professors and other colleagues of School of Economics, Prof. Sarveswara Rao Garu and Prof. Parthasarathy, distinguished guests and colleagues. This is the second in the series of Seminars organised by the School of Economics. The first one is on 'Births and Population Policy'and this is on deaths and suicides among cotton farmers in a gap of two months. Both are very important and of high economic and social concern. To be frank I know little about suicides but the concern has come in a large section of the society. Agriculturists are trapped into debt and found no other means to overcome but to leave the World. With this concern when I came to know about NABARD UNIT and Prof. R.M. Mohan Rao is undertaking a study I requested him to organise a Seminar on Suicides among Cotton Farmers. So let me thank first the Seminar Director for responding quickly and for organism this seminar. The present discussion is timely in the context of Economic Reforms and market failures and uncertainties. Hope the sessions that follow this inaugural address bring some points. Now I thank our Vice-Chancellor for his kind support all through inspite of his busy schedule after returning from the Conference of Vice-Chancellors of Common Wealth Universities.

Thank you Prof. Hanumantha Rao for your lucid address. I will be echoing what everyone here is speaking that we are particularly impressed by your great concern for our society. Many of us came from agricultural background. Your concern for institutional framework to meet such contingencies really require a thought and action. To stress a few, mere technological break through may not be sufficient but the responsibility of the Government in evolving institutional credit to small and marginal farmers, crop insurance, providing extension work from lab to land, strengthening of revenue system, deserve attention. I thank Prof. Parthasarathy for his concern to the suicide victims. He characterised it as an indication of a sick society and stressed the need to go into the causes of such sickness even in a sector like agriculture. I thank Prof. Sarveswara Rao for stressing on the need of social psychologists to analyse the institutional structure and power structure and examine the cultural factors to analyse the real issues. I thank Dr. Shivaji, former M.P. and Member, Central Cotton Advisory Board and particularly Sri Gutta Sambyya a cotton farmer from Prathipadu Mandal in Guntur district, Dr. A.Prasada Rao, Soil Scientist and Associate Professor, A.P. Agricultural University, Dr. K.J.S. Satya Sai, Assistant General Manager, NABARD, Mumbai and Sri Raghavendra Rao, Asst. General Manager, NABARD, Vizag district and other officers of SBI who came here sparing their valuable time to share their views. Lastly, I thank all the colleagues, guests, scholars, students, invitees and the press in particular, for making this seminar a success.

13

Suicides of Cotton Growers

Alternative Perspectives

RAPPORTEURS' REPORT

Prof. G. Parthasarathy

Among the alternative perspectives, liberalisation, indiscriminate use of pesticides, inadequate access to institutional credit, heavy debt to money-lenders ranks the highest. Prof. Hanumantha Rao this morning referred to the fact that small farmer is not averse to risk and small farmer has been taking more risk than even the big farmer. So risk is not to be avoided but there must be institutional effort to protect farmers against risks. All these issues can be discussed now. The issues are alternative perceptions regarding suicides. One perception is liberalisation has been the cause. There is a paper by Leela. Prof. Leela's paper brings out that liberalisation has not been the cause. Infact, liberalisation could help the cotton farmers by getting them high prices. Another perception is input adulteration and lack of extension. The pesticides have been adulterated and the third perception is risk. There is a heavy risk and insurance is necessary. All the institution's approach to the farmer has to be changed if necessary.

The discussion emphasises the need for institutional support to cotton farmers. Such support is needed in several areas : insurance against crop risk, access to institutional credit, etc. Such institutional support provides the conditions for minimising suicides. Among small farmers and tenants, the access to institutional support does not by itself end suicides. A large number of small farmers and tenants under similar circumstances did not commit suicides. There is a need for a deeper probe into the causes of suicides which may be related to atleast in part to the family circumstances. Certainly the erosion of value system in which value for life is considered sacrosanct should be considered a fundamental reason for suicides. The erosion of value system in rural society particularly among the farming community is a cause for concem.

Prof. K.S. Chalam

Chairman and friends. Though I am trained as an economist over the years, I have developed inter-disciplinary perspective under Prof. Sarveswara Rao. One important perspective that is missing has already been pointed by Prof Sarvesara Rao. I have with me a study conducted by Dr. Sudarshan Reddy, Dr. Venkateswara Rao on 50 families in which suicides were committed. One very interesting aspect is that majority of the deceased farmers belong to backward classes and I feel that none of them had a tradition of farming in their families. They are the new entrants into farming and more so cotton farming. They are, I consider, the drop-outs from the caste system, in the sense that, if he is a weaver and if he fails as a weaver he sells away whatever property, he has and gone in for cotton farming and therefore he has no community support which is generally available in our traditional system. Majority of the farmers from Guntur district are from a particular caste. I am not casting aspersions on any particular caste. This is a very interesting phenomenon that is emerging.

I know one or two families in Nalgonda and Warangal districts. Since Prof. Sarveswara Rao said if you take out a case study, this might throw some light on the whole problem. One weaver failed as a weaver, sold away all his property, bought land, the land belongs to a Reddy farmer and he sold away that land, gone to Hyderabad and he again came back to the same village as a trader, as a money-lender and he prompted the same farmer to take up cotton farming and prompted him to take loan from him and has given him the loan. He has also given him pesticides as a loan and ultimately this farmer failed and committed suicide. This is one case. Interestingly in this family, his father is still alive.

All those farmers who committed suicides are in the age group of 30-40 and the elders in the family are still alive. Only the persons belonging to this particular age group have committed suicides. Why this is happening? It has something to do with the transfer of land in the rural areas. Slowly, particularly in Telangana, the farmers are either becoming absentee landlords or selling away their land in the rural areas and go into cities and this land is bought mostly by the backward communities where the caste occupations have failed. They are shifting from their traditional occupations to farming. As a result, they do not have the necessary background, or community support, the economic support and may be the technological support in the particular case. I do not know how to relate this with the suicides. In the study they have brought out that 85 per cent of the suicides are committed among the backward classes, 10 per cent among scheduled castes and around 5 per cent among the upper castes. This is a very serious problem that is happening in the rural areas of Telangana and it might also shift to my area, North Andhra. I am told that Cotton is slowly entering North Andhra because of the delta farmers entering into these areas. In fact in response to Prof. Parthasarathy's article Revathi has brought out a very interesting observation. In Warangal district, the water sources are dwindling. They have come down more than 1 metre. But still the delta type of agriculture is coming into these semiarid zones and this is also one interesting aspect that needs to be studied. Here we have to understand the land transfers that are taking place and the inexperienced farmer whose traditional occupation is not farming is entering into commercial farming.

Prof. N. D. Raj

This is a complex issue having multi collinearity and one cannot delineate the psychological stress, the sociological bias and segregation, the liberalisation impact, the failure of the institutional support and a new dimension raised by our learned speaker, the Governmental negligence. Probably the Government was not aware of what steps to take up. So it has neglected. Now as he rightly put it we have to suggest the steps the Government should take up and I would confine myself to the first issue.

Now coming to Globalisation issue, suicides are not confined to India alone. They are all over the world. We do not have adequate data but we have the perspectives. Humanity is in a transition phase. In the transition phase new systems evolve, new relations evolve, new patterns evolve and in the process again we come to the jungle law—survival of the fittest and the death of the weak. So in the case of cotton farming it is the small and the marginal farmers and tenants, and again, as Prof. Chalam has mentioned, the socially backward communities like scheduled castes and the other backward communities. Why? The reason is it takes time to evolve shock absorbers. One shock absorber as suggested by Prof. Hanumantha Rao is insurance. Another shock absorber is price mechanism or market mechanism. Now in the process of Globalisation the existing systems and institutions are devastated. So it takes time to evolve new institutional factors and new systems and the guiding principle is profitability. Unless profitability is high new system would not come in. The same set of insurance will not benefit the farmer if he commits suicide. You have to have a different set of insurance system, or insurance package. Similarly, the same type of marketing system would not help in the liberalisation or Globalisation process. Because your market network is linked to the international network, international market system, international pricing mechanism. Before it is well linked and before the people at the grassroots level receive the benefits, there will be problems, issues, not necessarily death but what we call increase of immiserisation. Immiserisation is bound to come in the process of globalisation and liberalisation. Immiserisation will come quickly in the most vulnerable sections of the society, in the primary sector and in the primary sector again risk taking sector and among the risk taking sector the socially and economically backward communities. Globalisation cannot be stopped. We have to think of the institutional and other arrangements which will take care of the Globalisation process and solve the problems of the vulnerable sections.

Dr. Y. Shivaji

Without adding anything to the existing discussion I would like to elicit the opinion of the House, whether my observation is right or wrong because I visited almost all the families that committed suicide in the Guntur district. One of the observations is that we were discussing about the shock absorbers, insurance and other things. I observed that most of them are not from joint families. The so-called Hindu undivided family system collapsed. Had it been there, his brother, or mother or sister-inlaw or sister or some body else in the family could have absorbed and consoled the fate of the bereaved. Is it the absence of the Hindu undivided joint family one of the reasons for committing suicides in the absence of any consoling agency in the society?

Prof. C.H. Hanumantha Rao

Certainly as Dr. Shivaji says, it seems to be one of the factors. As he rightly said, had it been a joint family, then atleast, psychologically there will be people to console him and to give him some kind of encouragement and support. It would have made a very big difference. The point really is we cannot go back to joint family system. But, that does not mean the inference has no value. It has a tremendous value in terms of implications. Now there has to be new forms of institution of joint family type, new forms of collective support so that individuals as Prof. Sarveswara Rao has also mentioned the need for intervention so that farmer should be assured that there are people to take care of him and that the State is there or some institution is there. So this is a very important implication which arises.

Prof. V.S. Bose

Dr. Shivaji's observation about the occurrence of suicide in the nuclear family is very interesting. Committing a suicide is considered by Psychologist as a maladoptive response to crisis situation. It is a symptom of a depression. It has two types of implications. One, a clinical implication and two, a social implication. Clinically speaking if a person has a tendency to commit suicide or makes repeated attempts to commit suicide it is an indication that he is suffering either with a disorder called depression or with another disorder charactrised by delusions of persecution. The person fears that some people may do something harmful to him. As a reaction to that he may commit suicide to avoid that danger. This is basically a clinical picture. When it comes to social implications, when a person attempts to commit suicide there is some benefit that he hopes to gain. Though he wants to live his life there is some sort of a gain that may be obtained. The gain may be either financial or just recognition etc. A few people take that person as a role model and just follow him, sometimes in a mass way, when there is a whole spate of suicides. When a person commits suicide no matter whether it is a clinically or socially based, what is most important is he is subjected to stress. When we are under stress there are three ways of how a person perceives that stressful situation. One is as a threat. "Oh! this type of calamity has come to me and my family. I do not think I can help it and it is not really worth living". Another type of a reaction is showing an aggression towards the source of stress. Usually, the source of this are the people, the system and institutions that are responsible for this type of situation. So he takes some sort of a revenge against these people. These two types of reactions are negative reactions to the stressful situation. When the person develops this type of a thinking the chemistry of the brain changes. There are some toxic substances that are produced in the brain and those toxic substances ultimately lead to the development of a disorder, depression and as a result of that he commits suicide.

Why is it that the frequency of occurrence is not more in joint families, why more in nuclear families? Possibly, in joint families there is more opportunity for people to share their agony with others in the family. When they share their pain or when they cry clinically it is called catharsis. Catharsis is a therapeutic technique. The tears that we shed when we peel an onion when chemically analysed are different from the tears we shed when in grief. The latter contain some toxic substances that are secreted in the brain and are expelled through our tears. So, by sharing their grief with family members, these toxic substances that are accumulated in the brain are eliminated and the thoughts of committing suicide, are less likely to occur. There is a relief for the person and he is likely to think of facing the situation by taking it as a challenge and see what best he can do.

When a person takes that stressful situation as a challenge and then thinks "Let me see what best I can do under the circumstances that are prevailing now. Let me try to cope with it" then the chemical substances that are produced in the brain are another group of chemical substances called natural anesthetics and natural opiates. The characteristics of these chemicals are that they are natural painkillers and energizers. They give the person a lot of resistance and strengthen his immune system.

Prof. Chalam's observation that a particular age group seems to be more prone in those families to suicide could possibly be because this age group bears the responsibility of the family. The older generation were probably not taking the family responsibility on their shoulders because their children might have been taking it up. The person who holds the responsibility feels more stress, so their suicide is only a reaction to the stress they expressed due to crop failure.

Prof. Parthasarathy

Now the question raised by Dr. Hanumantha Rao is, quite pertinent. We cannot go back to joint families. You need to have alternative institutional support.

Prof. C. H. Hanumantha Rao

The analysis of stress situation, when an individual is alone and is in joint family is well taken and the point made by him is very valid. About the remedy, I would I like to raise a relevant question here. Had it been a joint family perhaps they would not have invested so much on pesticides also. Let us go to wider issues of resource allocation and decision-making in a joint family, we can study, if possible, if there are data on the joint family among small and marginal farmers, apart from nuclear family, I don't mean subjecting the data proving or disproving. Adventurism in the nuclear family is one thing in decision-making. In joint family the cropping pattern will perhaps be different and he will not be allowed to run away with so much of investment and even if some investment is made the decision will be a joint decision in the family. Nobody can blame a particular individual. So your analysis of stress is well taken. But behind this all, in a joint family, there are many other factors in decision-making and sharing responsibility and all that. This will go to soften the blow. Again those who have not committed suicide among small and marginal farmers, if you have a sample of those farmers, you may find large number of them are also backward classes. In fact, I wanted to make a general observation about the analysis based on this particular sample. We have selected those who have committed suicides and then analysed their characteristics. But if you have a random sample of small and marginal farmers in general and again where nuclear families as well as joint families then what kind of characteristics would come among those who are small and marginal but have not committed suicides? I expect, atleast in Telangana region to which you have mentioned, majority of them could be backward castes also. Now I am not discounting the data but your point is well taken about the support system not being available for farmers and among them for backward castes.

Dr. A. Prasada Rao

I have two or three points to make and may be one or two questions. I am not from a medical profession, I would like to know whether the depression is a disease by itself or a manifestation of a malady that is somewhere else? Coming to Dr. Shivaji's response, definitely the joint family offers the cushion compared to single family to absorb any risk. It was stated that the family members share the agony. It is not only the question of sharing the agony but they also share the burden. That is more important. After all, if you take the joint families of the rural system, pre-capitalistic feudal system which are in the transitory phase as it exists. See they always share the burden even if one member of the family goes, or if one family member loses heavily. The other family members still make up the loss. So they provide the cushion. So joint families definitely offer cushion and they will prevent suicides and there is no doubt about it. The other point I want to make is there are suicides even in high caste people also. Atleast, I have toured the districts of Warangal, Guntur and then Kurnool. In these three districts, as you know there are more number of deaths from backward communities and then SC families. The point is these backward families were working as agricultural labourers earlier traditionally. Their sons when they are educated are not prepared to go and work as agricultural labour. They are trying to take lands on lease and then they are pooling up investments from here and there and are borrowings mostly from private money-lenders. They have a weak social base, they have a weak economic base. Once they lose, they become heavily indebted and even if they have one or two acres of land to sell to discharge their debts, the market for land also collapses. When an area is affected by poor crop, the market for land, also collapses and then he will not have any wherewithal to discharge the liability for which the private money-lenders are forcing. At the same time, they have to meet family expenses also. So the main person who is bearing the burden of all this crisis feels that death is more pleasurable and honourable rather than continuing to live and face all kinds of social indignities. So the psychology of the person committing suicide must be understood from this angle. I think this point we have noted in our report also.

Prof. B. Sarveswara Rao

May I say just one or two words here. I was asking a medical friend of mine of what exactly is suicide? What are the types of suicides? It is very interesting to know from the medical scientists not a psychologist that there are three kinds of suicides. One is a person gets overwhelmed by a distress situation and he impulsively and immediately feels that there is no easy solution to this therefore he commits suicide. Second category of suicides, he told me is a very calculated suicide, rational suicide. He wants to hurt somebody or he wants to please somebody and there are many cases of that kind not necessarily agriculturists. We need not bother about that. The third case is, the depressing situation either for a short period or for long period goes on for 6 months or one year, it affects the functioning of the brain and psychological structures, as my friend Dr. Bose said. This goes on. Then he loses his capacity to think clearly, loss of clarity, of vision, of thinking. Therefore, he commits suicide. This is an interesting thing. Now how many of the farmers' suicides are in the first category.? How many of them are in the third category? Now my point is that the third category seems to be rather small, the first category seems to be very high. Impulsive, quick reaction, they felt that some how that nobody comes to their rescue—the Government, the community. There is no chance of recovering the losses and stabilise his position once again. That kind of an impulsive reaction perhaps has led to the suicide. The third case also is important, the depression case. It may be for 4-6 months, or 1 year approximately. Depression, unless it is given treatment, by a qualified medical man can lead to the disfunctioning of the brain and that means he loses his power of analysis, thinking and therefore sooner or later commits suicide.

Prof. B. Prasada Rao

I have one important point to make out in this connection. We have been cultivating commercial crops, like tobacco, cotton and several other types of commercial crops for a very long period. We have not noticed the suicides during the last 2-3 decades. Suicides occurred only in the last year. Actually, tobacco has been cultivated for a long time and there are people who have been incurring heavy losses in tobacco. If you look at the history, only 2 out of 5 years is successful years for them. Three years is a failure and two years is a success. But still they continue to grow tobacco and the same is the case with cotton. There are areas which have cultivated cotton especially in, say, Prakasam district, and they gave up cultivation of cotton and later on cotton cultivation has been extended to new areas. This has not happened in the past. Why it has happened only in the last one or two or three years? This is the one important problem. The second aspect, which of course, a friend has pointed out, is joint family system: the disintegration of the joint family system is one of the reasons for increasing suicides. I would say it is true that joint family system provides a sufficient safeguard but at the same time it depends on the type of the composition of the joint family. We have more people to absorb the risk. As the burden of the debt increases then probably the joint family system will work out. If we have more number of earning members, more of potential workers in such joint family, the risk can be absorbed very easily. But we have more dependents, more children and one or two workers are working in the family and the joint family system also cannot provide sufficient safeguards for such families. I think one has to look into the history of the cultivation of commercial crops and evolution of these commercial crops and this is an important aspect and we have to examine the background of the suicides in the recent past.

Prof. R.M. Mohan Rao

Before we go to another topic I wanted to give details on the social background of suicide victims. Coming to Dr. Chalam's point that most of the people belong to backward castes, there is no doubt about it. Around 81 per cent of the sample households consisted of backward castes and scheduled castes and backward caste is dominant. They have 6 to 7 years of experience in farming. They are not new entrants to farming; I am not saying six years is vast experience but I am only presenting the details. Then the other aspect is caste support. In about 70 per cent of the Warangal villages, the community's concern for other's misery is reported. The other aspect is control of caste groups over members. In this context, what I wanted to tell is recently, Lester David based on international comparisons has written an article in Indian *Journal of Criminology*. He pointed out that suicides are likely to be more in situations of high regulation and also of low regulation. My understanding of the data is, in Warangal suicides are high due to higher rate of social regulation because individuals do not want to deceive people and get away because of the social regulation. In Guntur as Dr. Shivaji has pointed out the size of the family is smaller compared to Warangal, of course, nuclear families there is no doubt about it. Because Guntur farmers are used to the ins and outs of the problems they have not lost heart compared to their counterparts in Warangal who felt a sense of social alienation.

Another thing is, we have also elicited information on the type of help they have received from parents., father-in-law, lineage kin, etc. I do not want to go into the details. The interaction, degree of help received from paternal, maternal and lineage kin is more in Warangal compared to Guntur where it is almost zero response to some of these question. So what I wanted to submit is, though we have not made any controlled study which we have realised a little later, Warangal and Guntur present two contrasting situations, one accounting for high rate of suicide in the total suicide cases, Guntur for relatively lower because 40 deaths in Guntur and 300 in Warangal and 87 per cent of suicide victims belong to forward caste in Guntur. With respect to samples in Warangal, first we located mandals with high concentration of suicides. and selected mandals to have a sample of 36. In Guntur there is no choice as only 36 suicide deaths are reported, and we have taken almost all the 36 in Guntur.

Prof. G. Parthasarathy

Now, one of the five conditions for suicides to be minimised is the presence of institutional support and structures for reducing the risk. If there is institutional support in terms of irrigation, credit, insurance the risk could be minimised. All these supports are necessary. One of the studies conducted on suicide deaths by a voluntary organisation states that deaths are due to lack of access to institutional credit. Most of the suicide deaths are attributed to the borrowings from the money-lenders or the local landlords.

Sri M. S. Sardesh Pande

A point is made about the institutional support. We have also made an impressionistic study We have not gone into a detailed study as Prof. Mohan Rao had done. As soon as suicides occurred we had undertaken just to find out whether the criticism, that institutional credit is inadequate and that is one of the major causes of suicides. We have undertaken some sort of a study just to find out to what extent we are able to take up the responsibility for the institutional failure. Our observations are, I am talking purely as a banker and that too from the view point of the SBI, I do not represent the entire banking system as such. We tried to understand and study this aspect. First, we have looked into the inadequacy of finance that is the first step we started. We wanted to find out whether the per acre finance provided to the farmer is adequate. We visited all the districts which have witnessed the suicide cases and tried to find out whether the scales of finance in every district for every crop are adequate. The banks jointly fix scales of finance for each crop identified for that particular year. It is an yearly exercise. We tried to find out whether the scales of finance that we have fixed for cotton crop in the district where this particular problem was observed was adequate. Our observation was that it was decided by the technical committee. In all the cases they had taken into consideration, the ground level factors and the prices of various commodities and inputs for growing cotton in that particular year and the scale of finance was considered to be adequate. Then we tried to find out whether the loan that was given by the banks to the farmers was given in consonance with the scale of finance and the acreage that was being grown over there. Our observation was that branch managers at the operating level were following the scales of finance. We tried to understand why the farmers consider the finance released at the bank as inadequate. Our observation was, as was brought out in some other papers also, that many of the farmers particularly in Warangal lease-in additional land and in many cases this lease is only for that particular year and was oral. Where the documentary evidence is not available to the banks and in the absence of such an evidence and in the absence of any laid down guidelines for the banks, this additional leased-in area is not financed by the banks. The farmer was not able to produce sufficient proof of his cultivable rights of the leased-in land. So automatically he gets insufficient financing for the total area of the crop whatever he is cultivating. So the farmers were not able to get institutional support to the required level. But here we have to make out the distinction of the lands that were traditionally owned , leased-in and cultivated by him for that particular year.

Secondly, we found, as I have said in the beginning, the scales of finance are fixed in each district depending on various inputs that go into the cultivation of particular crop, including the seeds, fertilisers, pesticides, labour components, then irrigation and everything. All those components are on the basis of technical advice given by the Department of Agriculture, that means the dosages for fertiliser, the number of spraying that are to be taken up by the farmer etc. Our observation is that in many cases the farmers in their anxiety sprayed the pesticides which are not relevant to the crop, as well as in overdoses. With the result there were manifold repercussions—one was the cost escalation which was not available to him from the bank finance. In the bank finance the built in cost was there as per the norms. Secondly, another dimension of this was in many cases because these sprayings were in excess of what is required, probably they were ineffective and the pests were developing some sort of resistance and thereby the damage continues to be there. Thirdly, for this additional cost on pesticides he had either resorted to money-lenders to borrow additionally or more easily he has gone to the suppliers or the dealers of pesticides and fertilisers. We read in a number of cases spurious pesticides were supplied to farmers. In addition to that, the pesticide dealer also advised the farmers on the variety and doses of pesticides. Here we have to mention that, some sort of technical support that is required to the farmers from the Department of Agriculture or from the extension agencies of the Government is absolutely lacking and the farmer does not have any advice on the nature of pests, the application of pesticides, variety, dosage etc.

Dr. Y. Shivaji

As far as Warangal district is concerned, my observation is that the institutional credit is almost dismal because I enquired the concerned Cooperative Central Bank about their level of finance. It was intimated to me that it was as low as 4.9 per cent for the entire farming community of the Warangal district. The reason being the Warangal Cooperative Central Bank is under default to the NABARD. Therefore, there is no re-finance available there. It is under section 11. There are seven such banks in Andhra Pradesh. I am very much afraid in the near future, their number may go up and they may be in majority. The second aspect is, not even 10 per cent of the rural credit required is covered by the institutions in Warangal district. Prof. Hanumantha Rao is mentioning about priority sector, priority sector covers not only agriculture but also several other aspects like small scale industries and self-employment groups.

Sri M.S. Sardesh Pande

As Dr. Shivaji says we also understand that for various reasons the cooperative banks are not able to get re-finance from the NABARD and they are not very effective in extending credit. We had conducted a study on our financing in Warangal district and other districts. Year after year, we have seen, it is normal and it has not gone down any time. What happens in the case of commercial banks, at least State Bank of India is crop loaning is one area where there is absolutely no hesitation. It very rarely happens, unlike other Governments sponsored schemes where it might happen.

Dr. Y. Shivaji

Suppose a small or marginal farmer goes to the bank and he owns say about 3 or 4 acres and he gets another leased-in land of about 10 acres. Is he eligible to draw loan to cultivate the entire piece because these 3 or 4 acres is good enough as a security and for 13-14 acres he may need about Rs.60,000 and what makes you to deny and what is coming in the way?

Sri M.S. Saredsh Pande

Basically we take the primary security as the hypothecation of the crop. When the farmer has no right to cultivate that land, he cannot hypothecate the crop that he is growing. So banker will not finance such leased-in land. I said in the beginning, it is one of the drawbacks. It is very much there. I agree. May be something can be found out. But, as on date, these are the guidelines that are available for the banks. For these leased-in lands we are not in a position to finance.

Dr. Y. Shivaji

I think there is some communication gap as for any economic activity upto Rs.25,000 no security is needed.

Sri M.S. Sardesh Pande

Rs.25,000 is the ceiling.

Dr. Y. Shivaji

No security is warranted and that 4 acres is worth one lakh rupees and what objects you ?

Prof. R.M. Mohan Rao

The problem is not with scales of finance. They are in the order of around Rs.7,500 per acre. The problem is the inability of the bank to confirm to the scales of finance in providing credit to the farmers. Another aspect which needs to be taken in fixing scales of finance and in preparing credit plans is while in the technical consultation meetings they were taking care of some of the escalations in the input costs there is hardly any effort to apportion the bank credit among different crops by taking into account the area under different major crops. If we take cotton in Warangal district, out of the total credit targeted, only 8 per cent is allocated for cotton, whereas cotton accounted for 42 per cent of the cropped area during the 1997-98 Kharif. I do not want to make any comments, these are facts. So in preparing future credit plans, it is not sufficient if we escalate scales of finance by a marginal increase in the input costs and others, one must also take care to see whether the credit flows equitably among different crops.

Sri D. Janakirama Raju

Sir, I would like to make a suggestion to the House. We all know that the scales of finance for each district is done by the District Level Technical Committee. The Agricultural University, has seven agro-climatic zones headed by a regional agricultural research station which has economists in their purview. None of the districts as on date has taken the economists' help in formulating the scales of finance. I would like to request the House to request the A.P.A.U. to involve their economists also so that it will be more scientific.

Dr. A. Prasada Rao

Sir, the problem in this particular case is, not the scale of finance. There is a ceiling/limit for farmers. The maximum in some districts is only Rs.20,000. I do not know about it in other districts. Atleast, in Warangal they have a ceiling limit of Rs.20,000/-

Dr. Y. Shivaji

The district level consultative committee sits once in 6 months for Kharif and Rabi. They sit and they decide the scales of finance. In that meeting all the concerned people including the Agricultural University or the Agricultural Department, Command Area Development Authority and other people are also involved. Some interested farmers are also eligible to present their case. All the representatives are invited but none of them will attend. That is the whole story and then they fix the scales of finance. What is happening at the ground level, I believe is that the banks are suffering for want of staff. The number of rural credit officers is less. There is no recruitment for the last one and a half decade. Therefore, in a rural area, suppose a bank caters to the needs of about 1,000 or 1,500 clients on each day, they are capable of meeting the demand of 20 to 25 accounts. In the peak season of 2 months or so, they are not able to cope up with the demand. It is happening in every district. The Government, day in and day out announces that the loans are rescheduled this and that. But, it is not happening at the ground level. So, I suggested to the SBI's, Regional Office, at V1jayawada also why not you recruit some 2 or 3 people to each rural area for the peak season of 2 months? They are not able to do that. This is the problem.

Mr. B. Janakirama Raju

I do partly agree with you Sir, but the crux of the problem lies on the otherside. Because, the farmers are not prepared to enhance their limits especially in our area, *i.e.* Guntur district. I also hail from Guntur district, although I am presently posted in Hyderabad. The farmers in Guntur district do not want to enhance their limits because they have to go around MRO's, MDO's and get papers and then go for mortgages. Instead of that they resort to gold loan. They take the gold loan across the counter. They get the money on the same day. The interest is less. Whenever he finds pests on the crop he immediately goes to the bank especially to the SBI. We have a system where our cash officer himself grants the loans on the same day. We will be delivering about 50 to 60 gold loans per day at the rate of interest that is given for agricultural crops. This system the farmer feels is better than enhancing his limit. That is one of the reasons. So what you said is partly correct, may be 20 to 40 per cent. The farmers are also reluctant to enhance their limits beyond Rs.25,000 because the rate of interest changes, and because of mortgage and other things.

Prof. C.H. Hanumantha Rao

From what you say, it is the inability to procure all these documents and things like that. Do not say reluctant. They will be very happy to take loans provided that support is there from the institutions and officials to provide all that.

Dr. Y. Shivaji

The Narasimham Committee stipulated that only small and marginal farmers are eligible, under the priority sector not all agriculturists. That is one aspect. The second aspect is, there is no monitoring agency in the Country and the average scales of finance is about Rs.775 per hectare if I remember correctly in the whole Country. A major portion of this goes for the adjustment of loans. The cooperative re-financing is always adjusted because, unless the particular cooperative society completed the collection and repays to the cooperative central bank, they are not eligible for re-financing. Therefore, they will adjust in the books and see that some loan they get from the central bank and it is redistributed.

One point of liberalisation occurs to my mind. I was mentioning to Prof. Sarveswara Rao what is happening for the last 7, 8 years? It is the same middle class which is responsible for any social change in the Country right from the freedom struggle to the Naxalite Movement. They are responsible to gather people or to propagate or to spend some money etc. Now, the same middle class is docile and they are mad with consumerism and they are purchasing various goods on instalment basis or something like that. They are indifferent. They are apathetic towards their surroundings. They are unconcerned about their kin and are more concerned about their nuclear family, his wife and children and that is one of the evils of liberalisation. I am correct or not, I do not know, we are aping the external world. We come across in Japan or Europe or America a single man's protest. A person at the airport or at the hotel says. "I protest against the nuclear bombs", "I protest against racism" etc. Here this is absent. That is the reason why for the last 7 to 8 years there is no strike among bank men, nor among the railway men, nor among the NGOs. What happened last year? Intermediate students are asked to take the examination, once again, after toiling in the mid-summer without any power supply. In May 13 lakh students were asked to take the examination again, and the first examination was cancelled. No little finger was lifted as a protest against in the entire State. The parents and children are jolly well taking the examination. But it is one of the maladies of liberalism thanks to Dr. Manmohan Singh and Mr. P.V. Narasimha Rao. I mentioned to both of them "this is the outcome of your liberalisation".

Prof. B. Sarveswar Rao

I just wanted to add to what Dr. Shivaji said about exchanging ideas. Infact, in a recent address I also said the same thing, the newly emerging middle class has not only become consumerist but has also become corrupt.

Sri K. J. S. Satya Sai

There is a lot of discussion about institutional credit to agriculture. No doubt there is some dilution in the priority sector lending over time. But especially in the first half of 90s there is a doubling of institutional credit to agriculture. May I now put this question? The delivery mechanism must have been spoilt during the recent times, as Professor Hanumantha Rao has mentioned. We are trying to improve the delivery mechanism. I still think that our efforts will be inadequate unless we bring the cooperatives and commercial banks into the fold. I think there is some diversion towards consumer loans and other types of loans. I think we are diverting towards new things like viability of cooperatives, viability of commercial banks, RRBs, etc., I think that is also partly diverting our attention from the delivery and institutions are increasingly finding it more risky to lend to agricultural sector and so they are shying away to make themselves more profitable. In one of the studies it was found that RRBs are investing in some investments rather than giving loans to become viable. But this brings to their notice that we are losing sight of the ultimate goal and we are trying to run after the viability rather than concentrating on the delivery to the needy.

One more thing, indebtedness is one of the factors claimed responsible for the suicides. We also sponsored another study in Maharastra, where cotton farmers have committed suicide Vaikunth Mehta Institute has conducted a study. We got draft results for that and in that, they covered two districts Amaravathi and Yawatmal. There 45 suicide cases were reported and all were covered by the Study and surprisingly it was found that 19 farmers did not take any loans, and only 26 farmers took the loans. Out of these, 26 farmers only 19 have reported suicides due to crop failures and resulting indebtedness and this indebtedness has been for the last 6,7 years. This analysis shows that less than 50 per cent of the farmers reporting suicides are devoid of institutional credit or any credit whether it is institutional or non-institutional. Given this, can we totally ascribe all suicides to indebtedness? This question comes again. Another thing is whether in 1987 also there were reported suicides in Guntur and

Prakasam districts due to crop failures, again involving cotton I want to know, whether crop failure in 1 or 2 years will mitigate suicide or is it a result of indebtedness over time and if it is crop failure only 1997 is the culprit. We have to think about it. As others pointed out, impulsiveness of a few farmers may not be a general trend. But, the Maharastra Study could not bring out clearly whether it is because of single years crop failure or is it accumulated losses overtime.

Prof. R. Radhakrishna

I would like to make a very brief observation. Now, in the context of liberalisation, the issue as raised by Dr. Shivaji, that the value system is changing and it cannot be reversed. There is a move towards more individualism and individual is guided by say interest and we have to accept these facts. We can not change the reality. The other reality today is, I am told that, the area under cotton has increased considerably this year, despite the suicides. So the farmer is also motivated. This motivation also you can not reverse it. The farmers credit share is very small. Even in coastal Andhra, in West Godavari, there was a study made three years ago. The share of banks in the credit needs of the farmer is less than about 30 or 40 per cent of the credit from the formal credit institutions. So you have to depend on informal credit markets. There is another study, done for Mahaboobnagar, which shows that 80 to 90 per cent of the credit needs were met from the informal sector. Now what are the changes that are taking place in the informal sector and as such the formal sector cannot reach the poor. Prof. Hanumantha Rao also mentioned this in the morning. Economists do not convince the formal sector to reach the farming community. How the vaccum has to be filled? Should one think of some innovations and creation of institution like Grameena Bank and similar alternative institutions? What type of institutions have to be created to protect the farmer? One of this is crop insurance, the one institution which we can think of now. Let us accept the realities. Let us accept market forces are penetrating. Let us accept that formal credit will not be motivated. It is very difficult to motivate them to meet the requirements of the farming community. In this context, what should be the alternative institutions we have to build? We should also have a discussion on that.

Dr. U. N. Murthy

I just wanted to raise one point here. With all respects to the elders in the Seminar, we should not allow this Seminar to become a kind of witch hunting by uninformed mode to jump from certain flimsy evidence to a what is called the blaming of the whole institution, that too a formal institution which has been helping the farmers in various other respects. The kind of logical gap is slightly troubling me. The second point is you are expecting a banker to finance a suicidee. A potential suicidee does not evaluate himself to remain in the world. How can you expect a banker to give him a loan on that self-evaluation is very important. Here is a situation where you are trying to say that the system is at fault, intermediate financial institution is at fault, the marriage institution is at fault, you are trying to pinpoint a kind of say an institution or the whole system the macro system at what level, the macro level, the micro level, the whole world level, and then you go into the philosophical level right back to the universal attitude which our Vice-Chancellor was suggesting. The jump I am unable to accept and I would like people to react on that.

Dr. K. Arunalakshmi

I am one of the few botanists here and I would like to give a biological view point of this problem. I know that this kind of pest problem occurs for many other crops. For example, in Indonesia there is a pest called 'Brown plant hopper'. The Government forced the farmers to rotate the crop. I mean not to raise it for two years and rotate the crop with something else so that they can eradicate this pest. One of the ways of eradicating the pest is just crop rotation. So I think soyabean is a very promising crop. The farmers in Warangal and Guntur districts were asked to rotate cotton and alternate it with Soybean.

Well, I just wanted to know whether the Government can interfere with this kind of problem and then force the farmers to rotate them and make the soil more rich, so that they can stop using pesticides or limit the use of pesticides. I think it is very important that the Government also should intervene and force the farmers not to grow cotton continuously for years and to make the soil fertile and this is very important.

Dr. Y. Shivaji

Time and again, all the representatives used to mention both inside the Assembly and outside, for the creation of Cotton Board and Chillies Board and little purpose is served because Tobbaco Board is there. There are 6 or 7 commodity boards in the Country like Tobacco or Spices or Tea etc., and our observation is that, some useful work is being done by the Tobacco Board. Unlike Tobacco Board, it is not possible with cotton, because cotton is an universal crop, tobacco is limited only to certain pockets in nine districts of Andhra Pradesh and in each district not more than 10 to 15 villages and the number of growers are also limited to about 75,000 or 80,000. About 45,000 farmers are there in the entire State. So it is very easy to regulate the crop in the case of tobacco. It is not that easy in the case of cotton. That is one aspect. The second aspect is with liberalisation, Government is withdrawing from every sector not only from the banking and we are demanding the Government once again to take the cart before the horse and it may not be useful at this juncture. Only for public consumption all these people will talk about the creation of commodity boards it is not going to solve the problem. Even the existing statutory boards concerned with the Ministry of Textiles, like Cotton Advisory Board or Textile Commissioner or the CCI and several other agencies, are not able to help the farmers. All of them are working against farmers and if some more agencies are created some more confusion will be added to the existing problem.

Prof. R.M. Mohan Rao

In some journals, articles and editorials creation of a board on the lines of Tobacco Board was suggested. We have made a study of Indian Tobacco Board from its inception till recently. One of the scholars has done it for his Ph.D. It has failed to draw the attention of the farmers and regulate area under saline soils. It failed to do anything in the larger interests of the tobacco growers. It has many other lapses.

Prof. G. Parthasarathy

We can now move on to another topic *i.e.*, adulteration of inputs, adulteration of pesticides and what needs to be done whether policy regulation is necessary or not?

Sri G. Sambayya

We also suffered a lot on account of cotton. Unable to pay to banks, money-lenders and even to labourers, fellow farmers used to stay at home and bolt the doors. In such situations we are afraid they may commit suicide and out of fear, for many days we maintained close vigil on them. Today, you are discussing more about those who have died. But, there are many others mentally upset, though not died. They have lost their lands and do not know what to do. Though, they are alive we all consider them as dead. I request you to reflect about such farmers also and see something good is done to them, so that they can live on their own.

Dr. Y. Shivaji

In your village instead of cotton, other crops are also being grown during the last four or five years. Can you tell us some thing about those crops like Gingelly, Soyabeen, etc.? How is the farmers position after they have switched over to other crops?

Sri G. Sambayya

After switching over to Soyabean, the position is better. Though, we had some problems in the initial years, during the last 5-6 years farmers have improved their lot. This year nearly one half of the farmers in our village have sown Soyabean. The yield also is 10 quintals per acre.

Prof. P. Leela

When I was going through the problem of cotton farmers I have come across a point of the raw cotton exports and the position relating to the trends in raw cotton exports, cotton yarn and cotton textiles. So in that context I came across the cotton trade of India and from that angle I looked at the problems of the cotton farmers. I found that in the context of the successful completion of the Uruguay Round of Multilateral Trade Negotiations and the agreement relating to textiles and clothing, I found that there is a lot of scope in the coming years for an expansion in cotton trade, raw cotton exports, cotton textiles and with that background and along with the liberalisation of the agricultural trade. Putting all these together, I feet there is a lot of scope for expansion in raw cotton export as well as a rise in the price for cotton. But at the same time we should also remember that trade itself will not help the cotton farmers unless the cotton textile producers who buy cotton from the farmers, and unless they take measures to modernise their mills to reduce the cost of production and to compete in the world market. This is the point I put forward in my paper. There is another point I have come across. Even though there is an opening up of cotton textile trade in the world market, the cotton textile producers in India are not up to the mark because they have not cared to invest their profits, ploughing back their profits, into modernisation of their mills or production or technique. So against this background, I felt that if there is a strong strategic thrust in improving the competitiveness of our cotton textiles by the traders as well as the producers, there is a considerable scope to expand our share in the world market and thereby indirectly help our farmers.

Prof. G. Parthasarathy

Prof. Leela feels that trade liberalisation will help increase cotton prices.

Prof. R.M. Mohan Rao

Recently I was informed by the former Director General of

Foreign Trade in India., Dr. P. L Sanjeeva Reddy, that we are only marginal exporters in cotton. One year we export and in the next year we import. There are also many other problems in augmenting our exports. A country must be a regular exporter and then only it becomes possible to pass on benefits to the farmers.

Dr. Y. Shivaji

The share of textiles in cotton is about Rs.30,000/- to Rs.40,000/ crores per annum.

Prof. C. H. Hanumantha Rao

I just wanted to supplement the point made about trade liberalisation in respect of cotton as Prof. Leela has mentioned. There is an article in the latest Economic & Political Weekly on the recent developments in the cotton textile industry. In fact, there is an editorial on this in the Economic Times also the other day. I have seen that article and it is a very big turn around in the last 4-5 years in terms of investment in cotton textiles and in terms of production, consumption, exports, of course all kinds of things. For the first time, it gives some kind of a hope that if cotton textile industry is modernised and is linked up with the whole liberalisation process. As Dr. Ashok Gulati has been making from his various studies that out of all the agricultural commodities, over a long period of time, cotton has been the most discriminated against in terms of prices—international prices and domestic, and next comes rice. Therefore as we have rightly been suggesting these two commodities will gain considerably from liberalisation. Cotton will gain most because unlike rice where the income elasticity of demand is very low, the moment we export rice the world prices will come down and the advantage will not be that much as in the case of cotton because of the high income elasticity of demand. As Prof. Leela rightly said there is a switch over to the cotton textiles now every where. That is why I was saying that the very same cotton which seems to have caused suicides can potentially become a source for the removal of poverty. Considering the experience of other countries and what the studies have projected, in the next and 10-15 years or so when this multi-fibre phasing out becomes effective, then it is quite possible that agricultural production in some countries may decline because people may migrate, agents may migrate to non-agricultural garment and textiles and this kind of a thing is coming up actually in a big way in some countries not only in India necessarily. But, there is no doubt that there can be a shift in cropping pattern, more and more resources to cotton in our own agriculture and less to other crops. So this is a kind of a scenario that we can expect in India also.

Dr. Y. Shivaji

I was studying about irrigation potential in Andhra Pradesh. I compared the data for 1988-89 and the present one 1997-98. I found that compared to a decade ago public sector irrigation potential, like canals and tank irrigation, went down drastically whereas it is compensated by farmers by tube well or wells. That is one aspect. The second aspect is minor irrigation is not dependable. Somehow or the other, for the last several years, agricultural sector has been given a complete go by. Only for public consumption we are talking all these things. But people sitting at the helm of affairs, administrators, planners and others say even the production went down by about 30 per cent. It went down in Andhra Pradesh and it is alarming. The rice production which was about 126.8 lakh tonnes during 1998-89, now its share is something like less than 90 lakh tonnes. So this is something alarming. All these things I hope we will discuss after lunch.

14

Rapporteurs' Report Technical Session-I

CHAIRMAN : G. PARTHASARATHY

RAPPORTEUR : C. S. MURTHY

The Chairman invited speakers to comment on the background paper on the suicides prepared by R.M. Mohan Rao for the NABARD Unit. The comments of the speakers can be grouped under 5 themes.

Is Caste a Factor in Suicides ?

It is noted that those who committed suicides were mostly B.Cs and S.Cs. Confirming this R.M. Mohan Rao noted that 87% of those who committed suicides were B.Cs and S.Cs. Caution should, however, be exercised while reading this figure because these caste groups constituted a large proportion of population. This view questions the validity of the proposition that social factors have a strong influence on suicides.

Is Inexperience in Cultivation a Factor in Suicides ?

There is the view that some people leave traditional occupations and take to the cultivation of cotton because people in the past prospered cultivating cotton. These people are generally new to cultivation. They are not quite equipped to cultivate the crop and when the crop failed they committed suicide. Speakers have drawn pointed attention to the need to examine the data on suicides from the point of view of the age of those committing suicides. The apprehension is that young farmers lacking experience in cotton cultivation suffered losses and therefore committed suicides. However, the evidence from NABARD Unit's Study shows that those who committed suicides were not particularly new to cotton cultivation. The suicides were claimed to be impulsive. Farmers were overwhelmed by crop failure and committed suicides.

Joint Family System as a Factor in Suicides

Several participants of the seminar spoke of the joint family system and its potential in preventing suicides. They spoke of the usefulness of the joint family system in absorbing the shock of crop failure. It is noted that stressful situations will be handled with little pain in joint families. A joint family helps to share agony Joint family situation is such that if one member of the family encounters loss/difficulty, other members help and mitigate the loss/difficulty.

No doubt joint family system cushions shocks. But it is a thing of the past. We cannot go back to the past. Joint family is no longer relevant. Small family is the norm now. Therefore other ways of absorbing the shock have to be found. What are the institutions that are relevant now? Many participants spoke of the relevance of crop insurance in absorbing the shock.

Institutional Credit and its Inadequacy

The participants spent considerable time on credit and related problems. It is noted that scales of finance were fixed by the bank after a careful thought. Inadequacy of credit may at times arise because the farmer is forced by bad weather into using pesticides many more times than recommended. Farmers also needed more money than they were granted because they were operating leased-in land. The bank cannot finance an unregistered lease.

The bankers' views were however contradicted. It is argued that credit supply was poor. Overdues constrained the banker to advance loans liberally. R.M. Mohan Rao noted that the problem lies with estimating the financial requirements. Area cropped under cotton is invariably more than planned by the bankers and this makes all the credit plans go wrong. Some felt that the inadequacy of the credit is because farmers diverted loans for purchase of consumer items. A word of caution is thrown in, however. It is important that the banker does not finance unviable propositions.

Crop Rotation to Reduce Crop Damage

The need to adopt crop rotation to overcome the problem of crop damage is emphasised. Soyabean is said to be the alternative crops. Government should intervene and force farmers to rotate crops. Farmers in the cotton belt, are now diversifying crops. They have found Soyabean to be profitable.

15

Suicides of Cotton Growers Related Issues

Prof. Jagadeeswar Rao

Colleagues and participants I welcome you to this second session. I would not call it the suicides of cotton farmers I would rather call it a discussion on problems of cotton farming and let us concentrate more on that and come out with suggestions, because suicide is a manifestation of a multiplicity of causes that have comeup in the wake of cotton farming. So let us not be constantly reminded for the unpleasant thing of suicides. Let us concentrate more on the problems that have come up in the wake of the changes in the cropping pattern in the growth of the area under cotton in Andhra Pradesh in general and in some selected districts. I think let us focus on that and come out with suggestions to solve these problems whereby you can prevent the unpleasant manifestation of the suicide. Before I ask the participants to present their views, I will make few observations. One important thing is all of you are familiar, and we must consider India has 80 lakh hectare under cotton cultivation. In Maharastra alone there are 10 lakh hectares under cotton and Andhra Pradesh is one where the area under cotton registered a phenomenal growth especially since the 80s. If you see the period 1980-81 to 1995-96 the area under food grains has declined and infact the area under food grains has registered a negative growth rate. But the area under cotton registered the highest growth rate of more than 5 per cent. This is the highest among all, the oilseeds, non-food grains, and commercial crops. In this respect also Guntur, Prakasam and Warangal districts have registered phenomenal increases. For example, in Guntur now 23 per cent of the gross cropped area is under cotton. In Prakasam 11 per cent of the gross cropped area is under cotton. In Warangal, in 1981-82 less than 1 per cent of the gross cropped area was under cotton. Now it is 16 per cent and 1 lakh hectares. So this terrible expansion and growth of area under cotton has brought in its wake certain problems and these are manifested in the unpleasant phenomena of suicides. So let us concentrate on problems that have come up in the wake of cotton farming and how to meet them. So I request some of the people who have not been able to participate in the morning and also others to participate in this session.

Dr. A. Prasada Rao

First of all, the Seminar as I understand set the objective of considering the suicides of farmers. So, I was wondering why the topic was confined only to cotton. The suicides of cotton farmers have to be viewed as manifestation of a general crisis and small farmers and tenants cultivating commercial crops are the worst affected.

AWARE's study has clearly shown that debts were incurred only to participate in agriculture and the diverted amount, if any, was negligible. The above study has also showed that adulterated seeds, pesticides etc. have been the major reason for suicides. Our distinguished chief guest has told that farmers are willing to take risk. But do the small farmers have the wherewithal to absorb the risk? So, unless the State intervenes to curb the malpractice of the market forces and introduce crop insurance and the like, we cannot retain the small farmer in agriculture. Next to small farmers, tenant farmers some of whom are educated and unwilling to hire themselves as agricultural labour, are able to pool up their limited resources with heavy dependence on private capital. Their problems are much more serious than those of land owning farmers. If one wants to have a sustainable growth in agriculture, the problems of tenant farmers have to be squarely addressed to.

Another important aspect is the heavy losses of the small and marginal in the product market. An exercise done by me on the minimum and maximum prices for various products in markets during 1997, shows that the small and marginal farmers, who sell their produce immediately after the harvest, on an average lose Rs. 105/- per quintal or 17% for rice, 27% in blackgram, 23% for cotton, 37% for chillies, 36% for garlic and 45% for onions. To avert such losses and help ensure legitimate incomes due to this class of farmers calls for State support notwithstanding what the W.T.O. objectives say or globalisation means. In brief, the State can not obligate its responsibility—the primary cause for distress. You look from any angle—either adulterated seed/ or fertilisers, or reduced credit or any thing have to be viewed in relation to liberalisation policy. Critics may say how suicides have become severe in Andhra Pradesh while liberallsation is applicable to all States in the Country. This is due to poor state of development in the State. We are the second highest users of fertilisers and topmost in pesticides-use in the Country and cost of cultivation of cotton is high in our Country and cost of cultivation of cotton is high in our State compared to others, everything is working against the farmer. Our irrigation area has come down due to reduced support to irrigation and poor maintenance of irrigation system. Introduction of capitalist mode of production without access to capital, attempts to use high level of technology without quality inputs and absence of skills for adoption of new technology are the major problems.

If you open up economy to exports, can we imagine that opening up farm economy will benefit farmers in view of heavy subsidy to agriculture, as in U.S.A. or European countries? In respect of cotton, the farmers are not able to get higher incomes even in years of higher production, as business lobbies connive to bring down the prices in the name of glut in international market etc. Thus, one should realize that distress, the farmers experience is not due to production problem alone, but also due to market-related problems. No market economy will survive unless market related problems are solved. In respect of fertilisers there is no infrastructure to supply micro-nutrients, zinc, iron and copper. Even for zinc and iron the farmers do not have any dependable source of supply. These inadequacies are built into input access and technological gaps. Dr. Chowdary and myself have made several recommendations keeping the stipulations and the obligations under WTO is a reality unless you abdicate from it. What we suggested was that an allocation of 10 of GNP added from agriculture must be given as a support to the agricultural system which is permitted by the WTO. Next, with regard to debt relief, without providing some sort of relief from private debt, indebtedness is not going to be solved. With the introduction of Kisan credit cards, which the government has agreed in principle and which the Andhra Bank has introduced a couple of days ago, we do not know to what extent the bank officials are serious.

Then, we wanted that the pesticides enactment should be strenghthened. Unless this is strengthened, you can not control the sale of spurious and substandard seeds and pesticides.

To strengthen extension and to bridge the technological gaps Government has incorporated many of the technical suggestions given by the experts. However, this policy is not backed up by financial allocation, with respect to strengthening research support. Both timely detection of secondary problems and rendering proper advice becomes critical in view of the location-specific nature of agriculture production. Our field observations in Pedagottipadu village in Guntur District proved what is successful in one village proved otherwise even in the neighbouring villages. This calls for not only supply of seed and the microbial culture, but also the market. It is difficult to sell the idea of alternative crops to farmers. Many of the agricultural practices like Integrated Nutrient Practices by the Agriculture Universities are also not sustainable due to lack of sustainability of recommendations. Inspite of globalisation, developed countries like U.S.A or European countries have not opened their economies freely under WTO's Clauses etc. India, also must find out ways and means to ensure self-reliance in agricultural production system to preserve national sovereignity. Countries like ours do not have adequate resources to import food grains in adequate quantities. African experience has shown that in situations of constraints to imports, the worst victims are the poor people and small farmers. If similar situations arise we will have to live with suicides. When our social consciences ceases we may have to content with the idea that suicides are also inevitable as part of the system. Any one with conscience will not allow this sort of system which calls for a sacrifice of lives of the poor.

Prof. Padmaja

Since morning we have been hearing about economic aspects of suicides *i.e.*, the role of economic bankruptcy. Now I would like to draw your attention to the environmental bankruptcy which is the consequence of excessive use of pesticides. Regarding environmental aspect, you know, in the soil there are a number of useful microbes. It is not out of love for microbes that we have to be concerned about them but because the beneficial micro organisms are useful to the crops. They are much more beneficial than the fertilisers. So we have to care much about them. The indiscriminate use of pesticides is not only responsible for economic losses but also for ecological losses by spoiling these micro flora. Once we loose our resources *i.e.*, soil micro flora, it is not possible to recover where as the economic losses etc., may be overcome to some extent by evolving suitable policies by the Government. Not only soil, in the environment also there are predators which are useful for the plants in the form of pollinators etc. Infact, they kill the bad pests which damage the crop. So we have to take into account all these aspects and I am happy that we are doing some work in the line of biological control agents as alternatives for chemical pesticides and chemical fertilisers. The question is how far it can solve the pest problem and serve as an alternative to the chemical pesticides there by solving the economic as well as the ecological and environmental problems. The concept is based on how to make the best use of what is present in nature. We are not creating any thing new. We are only exploiting the potential that is available on land. So if we use these bio-control strategies for insect pest management on cotton, we can minimise economic as well as ecological problems. From the Government side also we should have some support because the new strategy in the implementation stage *i.e.* in the transformation stage, it needs the cooperation of the policy makers as well as the users. In the case of chemical pesticides if one product is released into the market they spend so much by way of advertisement and they see that the product is sold. But this new concept of bio-control should be given proper enlightenment by educating the farmers, by way of conducting workshops, demonstration classes etc. To my knowledge the performance of the existing bio-control centres is not up to the mark. At present Agricultural Universities are taking up this aspect of extension and it should also be extended to conventional Universities doing research in this area. We have potential to take up the extension work by conducting workshops, seminars, demonstration classes, etc., we are using a fungus for killing the notorious pests of cotton. We have conducted field trails of the bio-control agent at Prathipadu of Guntur District in a small scale and we could get positive results and the same thing can be repeated on a large scale for which we need the cooperation of the Government by way of support to the R & D project and some bio-control centres must be ear marked for the Universities.

One point I want to say regarding the recent suicides in Warangal District, where Central Committee visited the particular affected areas. We have conducted a survey of the same areas and we could collect some of the dead insect pests that are responsible for causing major damage. We brought them to the lab and we could isolate some fungal strains from the dead insects. So we could establish that the cause of the death of the insects is not the pesticides but it is the fungus. So not only ineffectiveness of the pesticides, but the effectiveness of the fungus which is already there in the soil. So why can't we exploit it? For this we need to conduct further research. The insects that die out of the fungus can be easily identified in the field. The body becomes white in colour, with cottony coating due to the fungal out growth. Any farmer can say that the death is not due to the chemical insecticide in which case the body degenerates, becomes black with watery exudation, The same thing I have published in "Annadhatha" August '98 issue where I have elaborated how this fungus grows and how this can be exploited and how it can be taken to the commercial level. *Beauveria bassana* is the name of the fungus we have isolated which belongs to the class Duetewmycetes. What about its effect on other friendly organisms? That is a very important aspect because it should not affect the beneficiary organisms . To some extent the results are satisfactory. Relatively its effect is not, so dangerous as that of the chemical pesticides as the later's action is indiscriminate. So I want this body to look into this aspect of bio-control method for pest management on cotton. The reasons for suicides are largely economical. Why there should be economic reasons ? Because of crop failure. Why crop failure? Because of ineffectiveness of pesticides so there should be an alternative for the existing chemical pesticides. Somehow the farmer is raising loans either from money-lenders or from banks. He is not able to repay because of crop failure. So the reasons for crop failure and the reasons for ineffectiveness of pesticides should be examined and alternative strategies should be worked out.

Prof. G. Ramachandrudu

Basically I am not an agricultural economist. I am a demographer, currently showing some interest in health aspects, particularly health economics. When I was looking at the data that was given to me I found that the districts in which the incidents of suicides took place are the districts in which case fatality rate of three diseases, *viz.*, 1. Japanese encephalitis, 2. Gastroentrities, 3. Tuberculosis are high.

Logistic Regression Analysis is used in which the dependent variable can take only two values either zero or one and the explanatory variables are on right side. I will just list out what are the variables I have chosen and what are the results and interpretation. It is observed that there is a shift in Andhra Pradesh agriculture from food grain crop production to non-food grain crop production. The proportion is 73:27 in 70s *i.e.*, 73 per cent food grains and 27 per cent non-food grains in 1970s. Now it has almost reversed from 59 to 41. The proportion of gross cropped area under cotton has risen from 2.6 to 6.2. If we take other crops also that is what is happening. So I took the per centage of non-food grain area in gross cropped area as one variable. The second variable is per centage of cultivators having less than one hectare of operational land holding. I have three other variables, *viz.*, small, medium and large farmers. I tried one by one replacing one by another. I found that the small farmers to be is very significant when compared to the medium or big farmers. Third variable is the use of fertilisers per hectare. Fourth is the social development variable *i.e.*, rural literacy rate. Fifth is the rural crude death rate which measures the general health scenario and the sixth is the environmental factors related diseases case fatality rate. I got negative correlation between the probability of deaths and rural literacy rate and crude death rate. And all the remaining have got positive signs. The significant variables are the per centage of non-food crops in the gross cropped area, per centage of cultivators having operational land holding less than one acre *i.e.*, small farmers. Small farmers proportion increases the probability of suicides and then case fatality rate has got smallest standard error due to this. So the results show that the globalisation propaganda might have aroused false hopes on farmers and made them to do risky innovation. That resulted in heavy debts that led to the suicides. The interesting findings that environmental based illnesses might be on the rise because of over use of pesticides and fertilisers that might pollute particularly drinking water apart from air. These diseases are high in these districts that reported suicides. The findings of the study clearly focus towards inadequate and improper agro-reforms without building enough social and physical infrastructure in rural areas and adequate insurance for risk. It is bound to result in great strains on small farmers' families. The agricultural extension programmes that are discontinued by the Government should be revised and expanded for keeping farmers informed about changes in the economic and physical environment.

Prof. N.D. Raj

Regarding NPK, I have collected data in 2 villages of Anakapalle in 1978, for about 200 acres in each village. The data are there, and if you go again probably after the 20th, there will be a lot of change. Now pursuing the morning part about liberalisation, globalisation, *vis-a-vis* the counter reaction and the system failure.

On the other side, of the world what they are doing? Now here is a person in the place of Dr. Shivaji whenever we talk about environmental regulation this and that why should I listen only to Dr. Shivaji's views why not others? Others are keeping quiet? How can only one person fight against all these things on behalf of everybody? So what one needs to do is lobbying of cotton farmers.

Dr. A. Prasada Rao

I offer only two or three clarifications on the points raised. One point is, these liberalisation policies are made responsible for sucides of the farmers. They were existent even earlier. As student of agriculture the same problems which we are today discussing, starting from credit, fertilisers all those things they were there earlier. After the introduction of liberalisation policies they got intensified. The intensity is in the pace of intensifying the contradictions and then the intensification of aspirations. These two have to be related. That is very important. Intensification of the aspirations and intensification of imperfections in the system are responsible for the present malady.

Regarding the seed, as far as the cotton is concerned in Warangal district, only hybrid seeds are involved. For hybrid, for privatisation no certification is there. They are selling the seeds in the name of research hybrids. Actually no research component is involved. When once they are developed, when once the parents are identified, only production is involved. But the private agencies are selling the private hybrids in the name of research hybrid and varieties will not be there. In cotton varieties do not constitute even 1 per cent. As far as the food crops particularly rice varieties are important. In case of varieties the pure seeds, *i.e.*, breeder seed, nuclear seed are produced by the University and to some extent foundation seed also. Then it is the State Seed Development Corporation which organises the production of these foundation seeds and to an extent nuclear seed with the help of the university. There is a third seed certification agency involved. The varieties can be sold in the name of principal label. There need not be any certificate as far as varieties are concerned. It can be under truthful label. It depends on the firm which is selling. Suppose our university sells some seeds to the farmers there is no certification. They have belief in the university seeds so they take it. But what has happened. This facility is misused by the private agency. So this distinction *i.e.*, variety and hybrid being largely controlled by the private trade should be kept in mind. Regarding the bio-fertilisers, our limitations are, a lot of work is being done. I have to add only one point. As far as research is concerned there is no sole proprietory for any body We will be very much pleased to cooperate at any level. There is no problem at all in coordinating. We are very much open minded. Our University is very much involved in these bio-fertilisers but the research support that is needed is very modest at present. It has to be expanded. Then the limitations of these bio-pesticides are well known. Madam has told that the first step is the major problem. Mere recommending use of these conventional insecticides when the pest load increases beyond 50 is useless. These bio-pesticides were effective at a low level of pesticide load. Beyond that they do not work. In our report on distressed farmers, we have identified several extension gaps. In the recent state agricultural policy announced by the Government almost 90 per cent of the policy is trying to address to fill up this extension gap. But it has to be backed by funds. Unless funds are backing the policy at best the policy becomes intention.

Then the last point relates to price parity. It has been argued elsewhere that price parity between industry and agriculture products should be maintained. If you compare 1976, 77 and now, the price parity has come down to 80 per cent compared to 76, 77 around that period. It is well documented and also very much essential. Equally important is ensuring the income that is due to the farmer. I have presented data earlier showing the small farmer is losing anything from 16.7 to nearly 45 per cent of the income due to him. If we can provide some mechanism where the loss is returned back to the farmers definitely we can solve the distress to a large extent. But all these suggestions from whatever angle you see needs state intervention. But globalisation as a concept wants state to keep away and then give free hand to the market. Even WTO requires State intervention. It requires more and more of state intervention. So my strong plea is that Government or state in the broader sense should not abdi-cate from its responsibility to intervene and in favour of the farmers.

Prof. R.M. Mohan Rao

Farmers, particularly in Warangal are borrowing money at high rate of interest, to spray pesticides. Under the circumstances, they will not resort to indiscriminate spending on pesticides. If they are resorting to excessive use it is either because of the poor quality of the pesticides or non-availability of the required varities in the market. To consider their actions as irrational is not appropriate. Many farmers had considerable experience some of them are more knowledgeable than the extension functionaries at the field level.

Secondly, as Dr. Prasada Rao has said post-suicidal strategy of the Government laid priority to extension, In Warangal, I enquired the concerned in the Agricultural Department, whether any extension staff are being sent to refresher courses. I was given some details which do not convey much sense. I personally feel the agricultural extension staff who are merely administrator distributors of seed, pesticides etc, may not provide effective extension as they might have unlearned what they have learned during their student days, in the process of attending to jobs other than extenison. Further, it is reported that there are 48 generic varities of cotton. Can we expect the extension staff to deliver goods without refreshing their knowledge, in view of the multiplicity of seed varieties and mind bogging number of pesticides. It is difficult to restrict the number of varieties to the minumum as suggested by Dr. Prasada Rao because there is already a view, that free trade in cotton and restricting the number of generic varieties are not tenable. Even if we ignore such arguments, can the agricultural extension staff who are currently jackals of all trades and for all crops master the agronomy of so many varieties of cotton and varied pests is another dilemma to be resolved.

Thirdly, farmers are often blamed for not adopting integrated pests management. An examination of the agricultural production plans for Khariff 1997-98 in Warangal and Guntur district show that they have not stressed much on IPM. One of the components of IPM is biological pesticides. Where are they available for farmers to buy? Did any agency exist for the sale of such things? In the absence of availability of the biological pesticides in the market and lack of any effort on the part of the Govt. to provide them particularly in the two districts under study are we justified in accusing farmers for not adopting Integrated Pest Management?

It is only in the 1998-99 Agricultural Production Plan of Warangal, an effort is made to reflect on these biological pests. How can we expect farmers to use IPM in 1997-98 under the existing circumstances.

The signing of agreements in WTO without commitment to implement the terms of the agreement will not deliver goods. Governments will be signing on many agreements but not serious about implementing all the agreements. The available evidence in Warangal shows that agreement on IPM is not yet serisouly implemented. In the absence of implmentation of programmes at the ground level how do you expect attitudinal change in favour of IPM at the farmers level? One is at a loss to understand such statements.

Farmers are also accused of not resorting to rotation of crops considered as talisman for many ills plaguing cotton crop. We are all aware that the small and marginal are shifting to cotton at great risk to improve their lot. Can these farmers with one or two acres afford to resort to rotation and yet maintain themselves? Though the practice is good one in the case of the small and marginal farmer's it is not a feasible proposition.

The argument that the roots of the cotton crop go deep into the soil and nutrients in the upper cluster of the soil remain unutilised in the absence of clear rotation is also not very sound. At present, tractors are mostly used for ploughing and this helps mixing up the entire soil and as such the problem of under utilisation of nutrients in the soil does not arise. Probably certain types of nutrients not needed for cotton may remain in the soil. This can be overcome by applying such nutrients required for cotton alone.

Farmers are aware of the suggestions of above nature, the

benefits as well as losses. But they are not able to adopt for their own reasons particularly in the case of small and marginal due to the compulsion of limited acerage but not out of ignorance of short and long-term advantages of the various practices. The small have to be more concerned about the short-term strategy of their survival before; they visualise the long-term stratregies. This all of us must appreciate or understand.

The last observation I want to make is, we are not lacking in ideas and strategies. Our plans have plenty of these. But when it comes to action and implementation we are a great failure. This is what I felt after staying in Korea and observing the plans and implementation strategies of the two countries.

Dr. Bhaskara Reddy

Mohan Rao has explained many real things. In this context I want to say that though advanced research is taking place in University campus, we are almost alien to inter-disciplinary research. If we undertake inter-disciplinary research on suicides among cotton growers in collaboration with those from related departments like Botany it would be more rewarding, particularly, on bio-technology aspects. Problems are multi-dimentional in the nature and as such call for multi-disciplinary approach.

Dr. A. Arundhati

I would like to reflect on the Production of insect resistant transgenes as part of Integrated Pest Management. With the development of genetic engineering techniques in recent years, genes are identified from different Bacterial, plant sources for insect resistance. When once their mode of action on insects is understood, they can be successfully introduced into plants, so that they can produce their own insecticide.

Two sources of such materials has been used in Western Countries. One of them is BT toxins and the other is cow pea protease inhibitor. I am working in such area where we isolated a protease inhibitor from Artocarpus (locally available plant) which has potential to act as insecticide. Now, at National level BT toxins of bacterial origin are used as insecticide spray and attempts are going on to introduce BT toxin genes into different crops especially into cotton so that it produces its own insecticide.

The attempts made in this direction will not provide 100 per cent solution for the insect problem. It becomes part of Integrated Pest Management and also provide some solution, so that use of chemical pesticides can be reduced substantially.

Dr. Umadevi

Respected Chairman, I am from Botany Department and my commitment to this Seminar is that I am heavily funded by the Government of India for a project to investigate the possibility of using an organism which may have a potential as a bio-pesticide. The area that I am working is bio-pesticides. I was initiated into this project by a farmer with a scientific bent of mind but who has utter contempt for scientists. He is Mr. K. Ramesh from Guntur District. He was sure that using an organism that is a natural enemy of the insect pests and is found in the farms would be the ultimate and only solution to the insect pest problem in cotton. But when I read about that topic I realised that biological control can never be an alternative to chemical pesticides; it can only be an adjunct to pest management. This is the sad reflection I got after three years. Thinking that it will perform miracles when it has failed for 60 years abroad and 30 years here is utopic. That did not give me any impression that biological control and Vandana Shiva's article about environment and ecological friendliness will solve the problems of the farmers. Then what is it that I want to contribute in this field? What I found was that bio-pesticides failed as commercial ventures both abroad and even in our country and the reason is that the machinery that is used for bio-pesticides formulation etc., is costly. The bio-pesticides have a short shelf life. So they are very unpredictable; efficiency is lost very quickly. So I reflected on what should I do about it and the strategy I found is this. I have interacted with Krishi Vignan Kendra of Kovvuru in West Godavari district and one in Warangal district. BT and NPV are the two bio-pesticides that are available in the market. Both of them are unsuitable to cotton because one is sensitive to the pH of the leaves and other to sunlight. These are rarely supplied in a viable form from Coimbatore the only source in India. If I can familiarize the material that I am working with at a vermiculture level it would work. You know that vermiculture is practiced with ease by farmers. Krishi Vignan Kendras are playing an important role in this. Hats off to them. I really saw them training farmers in the 2 years that I went there especially to the Kovvuru people. They are familiarising vermiculture. I have seen farmers resorting to vermiculture in the fields. If we can train farmers to make their own bio-pesticides as an income giving as well as protective agent for their fields, biological control may have some role to play in Integrated Pest Management. We are trying to find materials which are cheap like bran and hay and suitable for growing the bio-control agent. Then we hope to educate and inspire farmers to take it up. It is not at an industrial level but just at a vermicultural level. It may help managing the pest problem to atleast 10 per cent. If I succeed in 2 per cent in this the fund given by the Government of India may be seen as having been spent judiciously. But let me tell you all, biological control is not a very successful method; bio-pesticides are spoken high in ecologically conscious society, but in reality they are very different.

Dr. Samba Murty

Mr. Chairman, and learned friends. There are several reasons for suicides of cotton farmers. It is possible to isolate the most important of these factors by adopting a comparative static approach. We have cotton farmers who experienced losses in 1997 and who committed suicides in large numbers and we have data on paddy farmers who experienced heavy losses in 1991 but did not commit suicide. What it is that explains this differential reaction? We examined if suicides are because of weather induced crop failure, Crop failed both in the paddy belt and the cotton belt and the crop loss is of the same order, about 2/3rd of the crop was lost both in the paddy belt and the cotton belt. Since the weather induced crop failure resulted in the same loss in both the regions and since it did not result in suicide in the paddy belt it would not have resulted in suicides in the cotton belt. From this, we can infer that weather induced crop failure is certainly not the factor responsible for suicide.

Next we examined if crop loss is the reason for suicides. If we look at the data, we find that the loss is about equal to the return that the paddy farmer would get in a normal year. In the cotton belt the loss is again equal to the return that the cotton farmer would get in a normal year. Losses are the same but suicides took place only in one place, in the cotton belt and these suicides therefore can not be attributed to losses.

We may examine poor extension and spurious seeds and pesticides which led to crop failure are responsible for suicides. Here we must understand that the farmers in the paddy belt are not particularly troubled by either of the problems. But in the cotton belt pests trouble the farmer. Spurious pesticides and fertilisers cause great deal of worry to him. In other words the cotton farmer is highly dependent on, he is at the mercy of the extension worker and the trader. These problems are specific to the cotton area. Poor extension and spurious seeds and pesticides must have been the reason for suicides. The farmer can not depend upon the Government to solve his problems. He can not solve these problems himself. Now the possibility of the Government resolving his problems in the foreseeable future is bleak and therefore, out of depression the cotton farmer may have killed himself because he has no one to turn to in this atmosphere. What is needed is we have to revamp extension machinery and we also need to device structures which help us to supply quality seeds and pesticides to the farmers.

Dr. N. Bhaskara Reddy

According to 1991 census about 78 per cent of the operational holdings belong to marginal and small farmers. Past empirical studies reveal that most of these farmers have low resource base and suffer from indebtedness. They are often illiterate. When they adopt new technology, they rely on the advice of traders when it comes to use of pesticides and fertilisers. This gives rise to several complications, such as excessive use of pesticides and fertilisers. This leads to environmental degradation but also to increase in the cost of cultivation. The recent suicides of cotton growers in A.P. happened mostly among these small and marginal farmers.

Several opinions have been expressed on the causes of suicides among cotton growing farmers in newspapers and academic journals in recent times. The causes are mainly low yields, crop damage, debt, untimely rains, supply of spurious pesticides and fertilisers and finally poor extension services. Profit potential lures the farmer to opt for cotton crop. Some farmers in villages like Gottipadu, sensing high risk and uncertainty conditions associated with cotton, shifted to soyabean, superior varieties of groundnut and pulses or by observing rotation of crops *viz*., blackgram with cotton, soyabean and cotton or pulses as a mixed cropping pattern. Thereby some of the farmers in Andhra Pradesh avoided losses from cotton. The individual farmer should prepare plans considering his resource base.

If the farmer felt that return from cotton crop is not remunerative he might have put his land to the cultivation of other commercial crops. The data show that the area under cotton crop in Andhra Pradesh increased over a period of time.

Along side this, the need for extension also must have increased. Extension agencies must cope with the increased demand for extension caused not only by the increase in the number of farmers growing cotton but also by the newer and newer varieties of pests that are damaging the crop. There is no indication that the extension agency has been able to squarely face this problem of increased demand for its services. There seems to have arisen a pressure on the services of extension agencies. Under the circumstances, it is likely that the poor cultivator is sidelined by the extension agencies. Because he has a poor resource base, his access to extension services gets greatly limited and therefore he is the most vulnerable of all cultivators. When technology is fast changing. and when pests and diseases attack the farm incessantly, it is the petty cultivator who is at great risk. Unless extension services are taken to the small farmer, he is likely to face crop failure often. Such crop failures as we have seen in the cotton belt will result in suicides.

To understand the latest development in crop production the State and Central governments should coordinate on this issue and should train its technical staff and farmers to face the challenges in crop production. There are two types of uncertainties, that is, nature made and the other manmade. The nature made uncertainties in crop production should be covered by insurance. The manmade uncertainty should be tackled by the efficient extension services and by strengthening existing cooperative societies.

Dr. M. Indira Devi

I agree with Dr. Prasada Rao. It is not a problem of Economics alone. It is a social problem. Suicide is a social problem which is more concerned to sociologists and psychologists and others too. I tried to explore the theoretical aspect of the suicides. Why farmers are committing suicide? What are the reasons? I tried to explore from the sociological theory. Since morning starting from our beloved Vice-Chancellor, failure of the family system is considered as very important. Families are becoming small. Dr. Samba Murty, Dr. Ramachandrudu and others have also pointed out that families are becoming small. So when families are becoming small, people are becoming more individualistic and family is not in a position to support the members in distress also. When they are in trouble kin ties are becoming weakened. Social ties, friendship groups, caste ties are also becoming weak. As some of you have pointed that farmers are migrating to distant areas to cultivate cotton and to get more money. So when they go to a new place they cannot have any social support. They are altogether new to that place. So some of the farmers who have migrated to other places also are facing this problem.

Most of the suicide victims are young farmers between 30 to 40 years in age and are mostly males. Here it must be noted that males readily commit suicides, females do not readily commit suicide. There is literature to support this hypothesis. Males can not express the psychic pain that they undergo in this situation. Then they totally isolate themselves in this anomic situation. They cannot express their emotional stress. So they do not see anything objectively That is why they die? Now-a-days young people from the small families are becoming more individualistic. People are becoming more individualistic and they are thinking that they can borrow from anybody and repay in next year. When it becomes a failure then they face misery. With more personal responsibilities they are becoming bold enough to take initiative but when they fail, they face this problem of suicides. They got totally isolated and alienated, and feel loss of status and as Dr. Samba Murty has pointed out a farmer who can not earn more money is looked down in society. All these lead to suicides.

Moreover in a society like ours, capitalism unleashes unregulated appetite of consumerism. Capitalists and Commoners alike, everyone wants to have everything provided there is money. How to get more money is the criteria. As cotton crop fetches more money farmers want to go in for commercial crop at all costs. So modernization process and rapid social and economic change is driving people to have unlimited aspirations. How to reduce them or minimise them? I think it is the responsibility of the family as an institution. Religion also should play an important role in the distress situation. I think it is failing in the modern society. People are not making use of the spiritual aspect of individuals life. That is being undermined. Everyone wants to have more money for what? Here economists I think health economists and other economists have to see how privatisation, New Economic Policy are driving people to distress. To illustrate people want to have better education. How much money we have to pay to have better education to our children? If they want to get a job how much they have to spend to the corrupt in the job market? This is a problem for everybody. This is also the case with privatisation of health care. Just if you have some accident how many thousands we have to shell down from our pockets. Because we are all earning in the Universities and having jobs we don't mind shelling down such amount but what about a farmer who cannot earn as much as we are earning here. So that is a problem. Education and health are two important variables for individual development. Privatisation of education, privatisation of health care, and other globalisation policies and liberalisation policies are affecting people in this way. Under these circumstances however, the small want to develop they can not develop. What is it the Government is doing if a small farmer wants to send his child to an English medium school not regional medium school? English medium school and speaking in English is considered prestigious by fathers who are not educated. They want their children to be better than themselves in terms of educational status. So where to get more money? How to get more money? The problem is the same not only in towns but in villages also. Both of them are more or less alike due to urban-rural interaction. They want to have good houses. Even in my own village, Prof. Mohan Rao's village, we have seen how it has improved today. Elevated houses, big buildings, all types of concrete structures in place of huts and mud houses. Houses are erected because of success in cotton cropping and they had a telephone in the village. They have every thing because cotton farming has brought more money to them. So when some of the farmers are successful in cotton farming even a small farmer also is tempted to invest to get more money, have better life for their future generation of children. So this is what is motivating and this is what I call unregulated appetite of consumers and capitalist alike. This is there in our system. That is why individuals when they fail in this cotton crop they are committing suicides. Sociologically speaking, because of social isolation, egoism, lack of integration and lack of normative regulation in society is also playing an important role in making people commit suicides. This is a fact. There are others who have pointed out about caste factor. May be some of the areas are dominated by some castes other areas may be dominated by forward castes. That may be the reason I do not know. I think the NABARD Report, as I feel personally is more exploratory. Lot of researches can be conducted by looking at the report. More hypothesis can be generated, more seminars can be held. Today Prof. Hanumantha Rao has been pointing out more and more seminars are required at various other levels to find solution to the problem. So this is what I wanted to present as a sociologist. I think our friend Dr. Bose, a psychologist will also reflect on these issues.

Prof. Bose

From the morning we have been discussing various causal factors of suicides. There is always cause and effect. Some farmers committed suicide. But we also have to remember, there were others who were under similar circumstances and yet did not commit suicide. Why is this so? Obviously this means that there are individual differences. There was a failure of the crop, whether it is attributed to liberalisation, pesticides, seed quality, or soil quality. This distress is there in the farmers. How did the individual react to distress? When an individual react to this distress by depression we call this reactive depression. This is different from the disorder of depression which is called endogenous depression. The student who commits suicide because he has failed in the examination has reactive depression. In reacting to a crisis situation like this, there are three modes of reaction. One is the impulsive reaction, where the person sees no way out of the situation and thinks that the only way out is suicide which is impulsively carried out. Some others react in totally different way. Initially, they react, adversely but slowly they build up ways of coping and readjust to the situation. They ultimately compromise with the situation. A third group of people keep suffering. They neither can take the decision to commit suicide nor can they learn to develop healthy coping mechanism and lead normally healthy lives.

Thus, the whole situation is to be looked at on par with natural calamities such as earthquakes, floods, etc. This morning, a farmer reported that "though we are alive and have not committed suicide we are almost dead for all practical purposes" They are like living dead and face a future of emptiness and are not able to come out of this belief and feeling. Just as the public and Government's reacts to such situations as earthquakes and floods and provides relief in different forms, there is a way how we can help these people who are under great stress. In addition to providing health services, insurance, medical and financial support, etc., which was highlighted by other speakers, providing emotional support is very important. There is something called a post-traumatic stress disorder (PTSD) syndrome which does not show up immediately but maybe revealed a few months after the crisis situation. For these people, emotional support becomes very important (perhaps sometimes more important than material support) to give life a meaning. This support can be provided by a network of counsellors, social service organisations and NGOs who can help these people to find meaning in life.

Prof. M. Jagadeswara Rao

I think all the relevant aspects have been pointed out by all those who made oral presentations, who presented papers and who made interventions. Now, I want to conclude this session and before that I wish to make my own observations on the subject.

As has been said since this morning, this is a very important problem on which Economists, Sociologists, Psychologists and others concerned with policy should reflect. A number of points have come up in the course of the discussion. In this context, I want to reflect on one of the important aspects associated with cotton cultivation that has come up again and again and that is the use of pesticides, the role of pesticides, and the kind of extension work that can be done with regard to the use of pesticides.

It seems cotton is an important crop which is highly vulnerable to pest attacks. This necessitates the use of pesticides and that explains why 55 per cent of the pesticides used in the Country are consumed by cotton crop alone. The cost of cultivation data for Andhra Pradesh reveal that pesticides constitute an important component of cost ranging from 25 to 40 per cent. In a State like Maharashtra, the per acre investment on pesticides is about Rs.2000 to Rs.2,500. So, pesticides is a very important cost component and this has to be borne in mind in considering the problems of cotton farming.

The State of Andhra Pradesh is a major consumer of pesticides in the Country next to Punjab, accounting for 28 per cent of the pesticides consumed. In Andhra Pradesh, two crops-cotton and paddy-together account for 50 per cent of the use of pesticides, the share of cotton being 40 per cent. So in Andhra Pradesh cotton is a very pesticide-intensive crop. Within Andhra Pradesh, Guntur and Prakasam districts account for an overwhelming share in pesticide consumption. Therefore, pesticide use is more in Andhra Prasdesh, more in some districts and in the case of cotton crop in these districts. The cotton farmers' suicides are in these districts.

There are some important aspects of the cotton crop which the agricultural scientists, agronomists and entomologists should bring to the attention of the cotton farmers.

1. This crop has 160-180 day cycle. The recommended spraying is 6 to 7 times during this cycle. But the tendency seems to be spraying the plant atleast 12 to 18 times.
2. The consequences of this are said to be very harmful. It seems that pests develop some kind of immunity at some stage in their 5 stage life cycle. Once they cross the first two stages in the life cycle, no chemical can kill them. This is one important thing that has to be brought to the attention of the farmers.
3. It seems in 1987-88 suicides by cotton farmers occurred in Guntur and Prakasam districts due to pest attacks. Surprisingly, the same kind of pest attacked cotton crop in 1997-98 in Warangal district. Nothing seems to have been learnt from the earlier experience and shared with cotton farmers in Warangal district. No control mechanisms appear to have been thought of in the light of the earlier experience.
4. An important thing that need to be stressed in this context is that farmers are rational and want to respond to the favourable market conditions by growing cotton on a larger areas. The best example in this context is the farmers' response to monopoly cotton procurement scheme in Maharashtra. In responding to the favourable market conditions and in their anxiety to take advantage of the opportunities available to them, cotton farmers appear to have forgotten some traditional methods of farming like crop rotation, which ensure soil fertility.
5. The rush for cotton cultivation, ignoring crop rotation has led to the emergence of a monoculture farm economy, especially in Warangal district. In 1981-82 jowar was an important crop in Warangal district, accounting for 26 per cent of the cropped area. Now it is only about 5 per cent. This has harmful effects on soil fertility.
6. Further, it seems that the roots of the cotton plant grow deep in the soil, absorbing minerals. Given this fact, if cotton crop is not followed by another crop whose roots spread only on the surface, the soil loses its fertility permanently. It therefore appears that cotton farmers do not appreciate

the virtues of crop rotation. Extension agencies have a role in this regard.

7. Above all, farmers have not been following modem methods like Integrated Pest Management (IPM). IPM is a very important concept relevant in this context. The Agro-Economic Research Centre, Andhra University was asked by the Ministry of Agriculture, Government of India to study the implementation and evaluation of the IPM in Andhra Pradesh. Similar studies are also proposed in Uttar Pradesh, Punjab and Tamil Nadu.

IPM is a system approach to manage crop pests.
The tools of IPM are :

— Chemicals (agro-chemicals, etc.)
— Biological
— Mechanical control (Removal of egg.masses, etc.)
— Cultural control (Crop rotation, pest monitoring)
— Genetic (Transgenic plant)

These are the different components of IPM. In fact WHO initiated this programme in 1980 itself. India is one of the seven countries who have undertaken this programme. Reduction of environmental pollution, achieving ecological balance and discouraging excessive use of pesticides are the important objectives of this programme. Indonesia implemented this programme earnestly and has been able to reduces losses in paddy crop. After 1986, it has banned 57 pesticides and withdrawn subsidies on pesticides. This has to be brought to the attention of farmers.

In India, as part of the IPM programme, Farmers' Field Schools are organised since 1994. Farmers and agricultural extension officers get trained in these schools. Those who get trained will go to their respective States and become 'Master Trainers' for agricultural extension officers and farmers.

These schools will be organised in villages where the use of pesticides is high. As part of our Study, we are asked to study the following:

(a) Whether the use of bio-chemical agents has increased and whether the use of chemicals has fallen;
(b) Whether there are attitudinal changes in farmers;
(c) Whether the quality of the crop has improved; and
(d) Whether farmers have got out of the clutches of the pesticides manufacturers/dealers.

Srikakulam district for paddy crop and Prakasam district for cotton crop have been selected. The Study will evaluate the impact of IPM in these two districts.

It is therefore important that this concept of IPM has to be propagated among the farmers. Farmers lack the knowledge about the type and method of application of pesticides. They may not be able to recognise the pest attack and the type of disease of their crop. There is also much confusion about the usage of the type and quality of pesticides to control pests.

This is because for the same pest attack on the same crop, some pesticides recommended by the manufacturers are not recommended by the agricultural universities and agricultural scientists. Further, there is a significant variation between the dosage recommended by the manufacturers and the agricultural scientists. Moreover, the dealer has the last say in the matter and the farmer has to follow him because of credit and other facilities provided by the former. The dealers' stranglehold on the farmers has to be broken.'

In view of these, number of measures arc necessary.

1. Farmers — especially cotton farmers — must be involved in this IPM programme. They should be made aware of and understand the issues involved.
2. The State Government should come forward and allot funds for the purpose.
3. Atleast one Farmers' Field School should be organised in a Mandal where pesticides use is very high.
4. Agricultural extension officers should actively participate in the process of making farmers follow IPM.
5. Farmers should be made to appreciate the long-term advantages of such a programme, rather than the short-term gains.

Some studies were made about the benefits of IPM. One such study relates to Pulladigunta village in the Guntur district. This study showed that the Cost-Benefit ratio in the case of IPM was 1 : 1.54. The cost benefit ratios for other measures like chemical control, biological control are not as high as this. Therefore, an integrated crop protection package has to be developed, which involves safe and judicious use of chemicals coupled with other control methods. Further, as Researchers from the Botany Department have pointed out, there are some insects in the nature which are farmer-friendly These should be allowed to increase or atleast prevented from extinction. The use of bio-pesticides should be encouraged. Farmers should be made aware of all these things. So, a great responsibility rests on the agricultural scientists and more extension work is called for. Farmers can not be blamed for some of the things they do as they lack education. Pesticides are used in large quantities in cotton cultivation and the suicides of the cotton farmers, in many cases are largely an account of this. I have concentrated attention on pesticides

16

Rapporteurs' Report Technical Session-II

CHAIRMAN : PROF. M. JAGADESWARA RAO

RAPPORTEURS : PROF. D.S. PRASAD AND DR. R. SUDARSANA RAO

Distinguished academicians from the departments of Economics, Applied Economics, Botany, Psychology and Sociology of Andhra University, Banking Personnel, persons from voluntary organisations and representatives of the farming community participated in the session. The session acquired an inter-disciplinary character by throwing on several economic and non-economic issues. The Chairman dwelt upon much in explaining the importance of the problem and advised the delegates to discuss about the root causes of the farmers' misery.

The Technical Session discussed the papers presented at the Seminar. For convenience the deliberations are grouped into following:

1. Sociological and psychological factors.
2. Environmental factors
3. Bio-control measures
4. Extension and crop insurance
5. Liberalisation policies
6. Governments role.

Sociological and Psychological Factors

A comparative analysis of cotton and paddy growers showed that crop failures, heavy indebtededness etc., are only myths. Poor and inadequate extension and lack of faith in the government to root out the ills in the near future have been basic reasons for the suicides of the farmers.

Issues like weak kinship relations, family system social and cultural changes in the rural areas, out migration of enterprising farmers were found to be dominant factors as far as suicides are concerned. Both the bereaved families and the surviving members need both emotional and meterial support. The suicides may have to be treated as a natural calamity. Social support system to contain impulsive relations of the farmers was advocated. It was also stressed that farmers need to organise themselves to have a better bargaining power in the input market and output market.

Environmental Factors

Attention was drawn to environmental bankruptcy. It was noted that microbes present in the soil are farmer-friendly and eco-friendly. Indiscriminate use of pesticides/ fertilisers is leading to extinction of microbes and loss of soil microbes is irreparable. Environment based illness has been on the rise because of the over use of pesticides and fertilsiers which is polluting water. It was pointed out that the suicides are high in those districts where the pesticides use is high.

Bio-control Measures

It was recommended that bio-control centres should be encouraged to educate farmers on how to combat pests by organising demonstration farms. Also, pesticide extension centres should be established in the Universities—both agricultural and traditional with proper coordination. Exploiting nature by the use of natural insecticides and use of bio-chemical strategies was recommended as it is better than the use of chemical pesticides.

Extension and Crop Insurance

Crop insurance through multiple risk insurance is not feasible was advocated. It was unanimously suggested that extension services are to be geared up as the present misery is man-made. Cotton is more prone to pests and diseases and proportion of cost of pesticides in the total is very high, to the tune of 25-40 per cent. Cotton accounts for around 40 per cent of total pesticides-use in India. Andhra Pradesh accounts for around 28 per cent of total consumption of pesticides in India. Even in Andhra Pradesh 2-3 crops only use pesticides intensively. As pesticides is a purchased input and its use involves a high proportion of cost in the total extension in the pesticides use is highly essential. It was observed that farmers committed suicides as they used heavy doses of pesticides for lack of extension. It was also observed that the suicides of cotton farmers during 1987-88 in Prakasam and Guntur districts were due to pest attacks on a wider scale similar to those appeared in 1997-98 and extension services failed to deliver goods.

Liberalisation Policies

Considerable discussion took place on liberalisation impact on suicides. However, there was general consensus that liberalisation policies have affected agricultural sector in the form of declining share of bank credit without any freedom to export farm products. In this regard, the need to bring cotton exports under OGL which goes a long way to help farmers getting higher prices is favoured.

Governments' Role

The Seminar has identified three areas in which State has to play an effective role. These are in relation to providing unadulterated seed, and pesticides, effective extension and marketing for ensuring remunerative prices to the farmers. With regard to first two aspects it was felt that government regulations has its own limitations. The seminar called for exploring alternatives to the existing channels in the form of roping in cooperatives, marketing societies in the distribution of inputs and encouraging NGOs into the area of extension. In the area of marketing it was felt that regulated markets also have their limitations. It called for introduction of electronic scales and greater access to market information to the farmers through the electronic media.

The Chairman of the session, as such opined that the Integrated Pest Management (IPM) has been ignored so far, though it was proposed as far back as in 1980. This programme needs to be implemented more intensively and vigorously. He has suggested that Farmers' Field Schools are to be established where farmers are given training by extension officers and agricultural scientists. Regarding the type of pesticide and the quantity to be used, it was felt that dealer is the ultimate person influencing the farmer, notwithstanding the recommendations of the extension officers and agricultural scientists.

Finally the Chairman suggested the use of bio-pesticides on larger scale. His observation is based on a study which revealed that wherever IPM is used yields increased by 20 to 30 per cent and cost of cultivation has considerably decreased.

17

Plenary Session

Prof. R.M. Mohan Rao

I request Dr. Y. Shivaji to preside over the plenary session. Dr. Shivaji is too well known to the educated elite as well as masses. He has been a crusader of the problem of Indian farming community in general and that of Andhra farmers in particular. He keeps himself abreast and upto date on farm problems and a most sought after person by farmers, policy makers as well as academics. At present, he is a member of the Central Advisory Board of Cotton. I had tough time in contacting him as he is busy visiting affected farmers, representing problems to the concerned policy makers as well as participating in Seminars, etc. But I pursued relentlessly as his presence will be enlighting to most of us, enrich the deliberations of the seminar and make our observations more reliable and practicable. He is the one who knows the pulse of the farmers and can reflect and represent their point of view on the problem more than many of us. I am happy he has kindly agreed to be with us. We are grateful to you sir for being with us and agreeing to preside over the important plenary session.

Dr. Shivaji

Dr. Mohan Rao and learned friends. Before going to cotton I am very much concerned about agricultural production in the state going down for the last one decade. The total cropped area during 1989-90 was 122.60 lakh hectares whereas in the current year 1997-98 it is as low as 97.83 lakh hectares.

Way back in 1984-85 it was as high as 12.16 lakhs tonnes and now went down to 4.52. In case of Bajra in 88-89 it was 1.97 and the corresponding figure of this year is 0.52. This is the same with other minor millets coarse grain etc. In case of Redgram it was 0.7 as against 0.65 and in case of Greengram it was 2.28 to 1.11 lakh metric tonnes and Blackgrarn 3.97 and 2.45 and total pulses being 6.94 to 5.34. The total food grain production in our state went down from 129.01 lakh metric tonnes to 103.10 lakh metric tonnes. In the case of Groundnuts the fall is steep, from 21.70 to 11.96 same with Sesamum 0.40 to 0.18 and same with Castor it was about 0.85 and now it is 0.16 lakh metric tonnes. Total oil seeds production went down from 23.10 to 15.09. This is true of mesta. and other crops. The point I wanted to drive is that every member in the Parliament or in the Legislative Assembly are asked to fill up a form indicating their bio-data. In that a column is indicated about the interest background etc., and invariably 90 percent of the members write on it that they are from the agriculture family and their main vocation is agriculture. Nowhere else in the entire world, such a good number of representatives from the elected body are from agricultural families. But nowhere else is agriculture subjected to so much of hardships as it is happening in our country. So it is to be taken concerned of.

In case of irrigation, during 1988-99 canal irrigation was catering to the needs of 18.69 lakh hectares whereas as per the Government figures of 1995-96 it is as low as 15.39 lakh hectares. Same is the case with tank irrigation which was 11.04 in 1988-89 and now the latest corresponding figures is 7.47. This shows that the state supply of water for irrigation purposes in the State went down from 29.73 lakhs to 22.86 lakh hectares. It is being compensated to some extent by the enterprising farming community of the State with their own tube-wells and erecting a motor and getting a power connection and paying for it.

We are discussing this morning about the cotton. I also got educated and got experience but I feel we missed the meaning of the problem. Everybody was mentioning about liberalisation and globalisation and its effects on the agricultural sector. But to my knowledge the liberalisation and globalisation is yet to touch the agricultural sector and rural sector. It did not percolate beyond Visakhapatnam, beyond Vijayawada and to any rural part or agricultural sector. A question may be asked as to why agriculture is not being liberated? Agriculture is still unliberated and it is still with shackles and chains. You take for example, suppose Prof. Mohan Rao is producing paddy in his own village I do not know the unit value of paddy in Devarapalli. I think to my knowledge, it is around Rs.330 or Rs. 340 for a bag of 75 kg. It is common knowledge that two bags of 75 kg. paddy is milled in a small rice mill or hull the paddy yields about 100 kg. of rice. The ruling price in the market for paddy of 2 bags is to a tune of Rs. 660 or Rs. 680. That means that it should not cost more than Rs.6.80 per kg. But the same rice is being sold at Visakhapatnam, Vijayawada or Tenali at the rate of not less than Rs. 12 and it may go upto Rs. 16. There are several instances where the ruling price of rice in the market in Andhra Pradesh is the highest in the entire country. We out beat even our counter parts of Bombay or Delhi. They are getting at a much lesser price. We are producing about 90 lakh metric tonnes of rice and the marketable surplus being about 70 lakh tonnes. But unfortunately the system is yielding to the pressure of the rice millers. The rice fields are there from time immemorial in the delta area for about 130, 140 years in Godavari and Krishna delta area but the rice millers failed to develop some infrastructure like godowns or bank guarantees etc., and the procurement price of rice is slightly weakened against paddy. Therefore the miller always finds it convenient to supply rice to the Food Corporation of India (FCI) whether FCI or Indian Government in deficit or warrants it. But, we are supplying 45 lakh metric tonnes to the Food Corporation of India. You take the example of West Bengal and Andhra Pradesh. They are one and the same as far as the population is concerned. In both, basically the people are rice eating people. In West Bengal, they do not contribute anything to the Central pool though they are producing about 128 lakh metric tonnes of rice, Further, they draw about 12 to 13 lakh metric tonnes from the Central pool. I asked the question what is happening in West Bengal? and how is it that we are contributing so much quantity to the Food Corporation of India? The supplied rice from West Bengal is being smuggled to Bangladesh. We are not producing any wheat but West Bengal produces about 10 to 12 metric tonnes of wheat. Same is the picture with wheat also. They do not contribute any thing to the Central pool from their wheat production but yet they draw another 10 lakh tonnes from the Central pool. Thereby the farmers in Andhra Pradesh get a lesser price for his produce, Paddy and the consumer is compelled to bear higher price. Therefore, we are pleading to the Government that the farmer may be liberated, that he may be allowed to mill his paddy, to sell it or to transport it to any part of the country or elsewhere. You may recollect, that for the last 2 or 3 months the price of paddy has gone up a little higher compared to earlier. It is because Pepsi Company exported 6 lakh metric tonnes of rice from Kakinada port also. When there was a strike in Kakinada port, the price of paddy went down and when there was a cyclone at Kandla port also the price of paddy went down. So we are pleading with the Government, several Finance Ministers from Dr. Manmohan Singh to Mr. Yashwant Sinha including Mr. Chidambaram. Day in and day out in their budget proposals they are mentioning about the freedom that the farmers are enjoying in this country that they can export or import or that they can transport or bring the rice for marketing. But in practice it is not being done.

Second aspect we are discussing is about cotton and I don't know whether it is brought to your notice or not that three kg. of raw cotton when it is ginned it produces about 2 kg. of seed and one kg. of lint *i.e.*, the cotton without seed but one kg. of lint produces about 8 metres of cloth. The cotton produced in the Nagarjuna Sagar command area is more superior in quality compared to the Egyptian or Sudan cotton. Cloth is being sold in the market at the rate of Rs. 120 to Rs. 150 for each metre. Therefore, for 8 metres of cloth the minimum realisation is Rs. 960 and the poor farmer suppling the raw material gets for his 3 kg. produce Rs.3 × 20 = Rs.60 and the cost of the seed is Rs. 10 for 2 kg. So the poor man gets about Rs. 50 and the rest of the amount Rs. 910 is going to other areas like the ginning mills or weaving mills or the combined textile mills or to the advertisement media or to the Government in the form of hawala or something like that or to the excise or to the exchequer it goes. The so called white gold is the harbinger of prosperity to all but not to the grower and this fact is to be born in mind. Way back about 25 years ago or quarter century ago the story of cotton started somewhere during 1971 and 72 in Guntur and Prakasam districts. I was told that at that point of time during '72 or '73 some cotton growers from Gujarat were on a pilgrimage to down South for the holy places like temple towns of Tirupathi, Madurai or Trivandrum etc. On their way to attend to their nature calls the bus was stopped in a nearby village besides cotton field. After seeing the growth of cotton there, they enquired the farmers since how long they are growing cotton in this area. The farmer readily replied that it is only for the last couple of years that we are growing cotton. Immediately the Gujarat farmers commented that the same cotton that destroyed all the cotton growers in our State is brought to you and she does not leave you until you are also destroyed. So therefore, their prophecy became correct within 15 years. During 1986-87 around 50 to 60 cotton growers in both Prakasam and Guntur districts, committed suicides and several committees several discussions took place both inside and outside the Parliament and a high power committee headed by P.D. Ojha the then Deputy Governor of Reserve Bank of India visited those two districts accompanied by the then Chairman of Andhra Bank Mr. Naik and the then Chairman of Syndicate Bank Mr. Malya because both these banks happened to be the lead banks in Guntur and Prakasam districts respectively They submitted a report to the Government suggesting certain long-term measures and certain short-term measures. Unfortunately all those suggestions were completely given a go by. None of them are being implemented. I congratulate and complement the Agricultural University of Andhra Pradesh because they worked hard. Dr.Appa Rao and Dr. Subba Rao the present Vice-Chancellor and others connected with cotton crop and the soil scientists, the agronomists and entomologists all of them worked very well and we could arrest the spread of white fly. At that point of time it was controlled. But the same tragedy is being repeated in Warangal and Guntur districts with more vigour and virulence and the total number of bereaved families is more than 300 or so. The situation in Warangal is similar to the tragedy that took place in Prakasam and Guntur districts about a decade back. Unfortunately at this point of time the agriculture department is non existing and they are with sound sleep. I have yet to find any agriculture officer in a village. All of them are residing in the neighbouring towns with the compulsion of the education of their children or comforts of town life. They are mad after chasing goods and everybody wants to purchase a TV or a scooter or some other luxuries. To meet that demand they are acting hand in glove with the pesticide dealers and the spurious seed manufactures and their sellers.

I was in Punjab during last week. I spent a whole day in the Agricultural University of Punjab at Ludhiana and I asked them how is it going in your State? To my surprise way back in 1960's Andhra Pradesh was well ahead of Punjab. During those days they were bringing water with the help of a camel and the wheel and now to my utter delight I could not come across a bullock cart in the entire Punjab State. There are 8 lakh farm holdings in Punjab for which there are 3.5 lakh tractors and 2.5 lakh power tillers and about 9 lakh tube-wells and almost every inch of the land is being irrigated and supply of power and water is free to the farmers and every village is connected with pucca permanent tar roads. During 1950-51 Punjab's total production of food grains was as low as 2 million and now their production went up to 22 million metric tonnes and the storage capacity of Government owned godowns in Punjab was 150 lakh tonnes. So, every farmer is provided with, some shelter to keep his produce and then to enhance or strengthen his holding capacity. Whenever the market is favourable to them they can part with their produce and there is no compulsion to the farmer to drive him to the market in Punjab. That is the situation there and you know pretty well that the geographical area of Andhra Pradesh is about 2.75 lakh square kms. If I remember correctly Punjab has 52,000 square kms. There is an Agricultural University in Andhra Pradesh catering to the needs of the entire state as in Punjab. But the annual budget of Punjab Agricultural University is billion that is 100 crores of rupees, whereas the budget of our Agricultural University is 80 crore both plan and non-plan expenditure put together. I was told by Mr. Pratap Singh Kairon who was the Chief Minister of Punjab that the authorities of agricultural university (when he was visiting the university at Ludhiana) pleaded for sanction of grants worth around 1.35 crores during his visit to the University. Immediately Kairon replied why not you take 2 crores of rupees? Why is this piece meal affair of 1.35 crores. It was a herculean task to the Vice-Chancellor and other departmental heads to spend 2 crores of rupees during those days. Unless that money is spent, the Controller and Auditor General will find fault. So they were nervous and shy. Immediately Kairon asked the Vice-Chancellor and the Registrar to appear before him on the next day at Chandigarh at the capital of Punjab and the Finance Secretary was called in, Finance Secretary raised some objections that he could not make such a huge amount available to the University. Therefore, he warned the Secretary of Finance, I will fire you unless you issue the cheque for 2 crores by this evening and the amount was released. So what I mean to say is that whatever party is in office in Punjab every Government is committed to the farming community. That's how it is going on. Now in Andhra Pradesh you are seeing what is happening here and it needs no explanation.

About liberalisation, I would like to add that we missed a point since morning. What is happening here is that cotton is one of the earliest commodities that is traded both nationally and internationally. Until 18th century cotton was used only by the indigenous textile industry in the country, and therefore textiles were exported from India to other parts of the world. But with the Industrial Revolution and the invention of the ginning mill in USA, cotton was available in America, Europe, UK and other parts of the world. They have found some alternative from India. Therefore, cotton was being imported from other parts but with the civil war in USA and other reasons and with the establishment of colonial power in India the indigenous industry was slowly ruined and thereby the export of cotton started. With the partition of 1947 the country lost more land suitable for cotton production. Around 3.23 million acres of irrigated land, producing long staple cotton went to Pakistan. Though it was 22 per cent of the cotton growing area it was contributing to about 38 per cent of the total production of the undivided India. Therefore, at that point of time, the production of cotton went down with the partition and about 85 per cent of the textile mills remained in India. To meet that demand there was a ban on export of cotton and exports are restricted and there was free import of cotton *i.e.*, import of cotton was being allowed under Open General License and exports were restricted. It was incorporated during 1948-49. There were some climatic changes too in 1948-49 thereby we imported lot more cotton from Pakistan. When Pakistan refused to devalue their rupee during 1949-50, the import of cotton from Pakistan was affected but the restriction still continued. Now we reached a stage that we are not only self-sufficient in production of cotton, but also exporting it to other parts of the world. Unfortunately the restrictions that were imposed during 1948 still continue even for the last half century. What I am doing is pleading with the Government for umpteen number of times that cotton is not liberated, despite liberalisation policies. If restriction on export of cotton are removed I am sure that prices will definitely go up at least by Rs. 5 to 6 per kg. A survey was conducted and it was found that for the last 20 years at least for 17 years the international prices are always higher by 15 per cent when compared to the domestic prices. Thus, the cotton grower is subjugated to the textile industry in this country. But unfortunately the Government is always playing to the tune of the textile lobby and always they see that the textile lobby is benefited at the cost of the cotton grower. Therefore, we must pass a resolution and that even the Assembly also went on record by passing an anonymous resolution and appealed to the Indian Government that cotton should be included for export under Open General License without any restrictions what so ever as is done in the case of imports also.

You were mentioning about the pesticides, Integrated Pest Management and other aspects I found in Punjab, that the contamination of substandard pesticide use is less. The main reason being, in Kharif they grow paddy and for Rabi they go for wheat. So therefore, the crop that is there in the fields whether paddy or in the Kharif cotton it does not last in the field for more than 6 months or so immediately they go for the next crop Rabi *i.e.*, wheat. Therefore the remaining ratoon crop may not harbour the pests. Even the synthetic perethroids are not used in Punjab. They are banned in most of the countries developed and developing. But unfortunately India is used as a dumping ground for banned pesticides and especially Andhra Pradesh. Prof. Mohan Rao was mentioning about the incidence of use of pesticides on cotton in Andhra Pradesh and in Guntur district in particular. It is no surprise that we are using about 22 sprayings for cotton in Prakasam, Guntur and other parts. Cotton accounts for only 5 per cent of the total cropped area on the country, yet it consumes about 50 to 55 per cent of the total pesticides in the country. So I suggested to the Chief Minister and he agreed to though it is yet to be realised. My suggestion is that the Government open a depot in each district or at least in Guntur district as a pilot project. The pesticides manufacturer, whether it is a Shah Wallace Company, Rallies or whoever wants to sell his produce or market his produce, supply and unload his produce in the Government depot. Let the Government put a seal on each packet indicating the quality or ensuring the quality. Then only the dealers will lift from the godown. If it is done, the Government also may fix some responsibilities on the agriculture department. The quality seed even if it costs something like Rs. 5 or Rs. 6 the farmers are prepared to pay a little more. Dr. Prasada Rao gave me a pamphlet indicating how a scam is going on in case of pesticides. For example, for the Rogar supplied by the Rallies India Limited the listed price is Rs. 253 per liter where as the sale price is Rs. 200 and the dealer price is Rs. 185. There is a wide range of difference between these three prices. Same is the case with Nuvocron supplied by the Ceeba. The maximum retail price is Rs.418 and sale price is Rs. 320 and dealer price is Rs. 295. So I would like to suggest to the Government and you can pass a resolution to open a depot for the pesticides and they are jolly well doing for the India Manufactured Foreign Liquor (IMFL). The entire quantity of pesticide that is proposed to be marketed in a particular district may be unloaded in the Government depot and it can be delivered only after the quality seal of the Government department. If Government wants Rs. 5 or 6 extra per each kg., towards cost of service the money can be utilised from the market cess or some other source.

Another point I would like to bring to your notice is that we are collecting about Rs. 300 crores each year as market and rural development cess. These are intended to develop some infrastructure for the marketing of agricultural produce. But unfortunately the entire proceeds, go to the Government treasury under general fund. In Punjab the entire market cess amounting to Rs. 200 crores is being spent only to develop the link roads in the villages. So these are some of my observations. My proposal is not that the Government should step into the market directly. The agricultural market committees are intended to develop the marketing system in the state. But this is not done properly.

Another suggestion is, for example, if Rallies Company wants to sell about 1,000 bottles in Guntur district, these 1,000 bottles are to be unloaded in the Government depot. From there the dealers will buy it. There after the agricultural department gives quality seed. In turn, the dealer is responsible to pay the money to the company. The Rallies Company is not supposed to supply the product directly to the dealer. It is just like in the case of IMFL. So that is one proposal.

Another aspect I would like to bring to your notice and that of Dr. Prasada Rao is in Punjab a peculiar virus has spread from Pakistan. Now it has spread not only to Punjab but also to neighbouring, Rajasthan and I am very much afraid that it may shallow Madhya Pradesh and Andhra Pradesh also in the near future unless some preventive measures are taken. I hope that Andhra Pradesh Agricultural University can take some necessary measures to prevent it. These are some of my observations. About 1½ years back, *i.e.*, during 1996 June I believe when elections were on; and when there is some agitation in Guntur district the Government of India allowed export of about 5 lakh bales of cotton. The bureaucracy in the Ministry of Textiles, sought some clearnance from the Election Commission. You know the Chief Election Commissioner Mr. T.N. Seshan came in the way and said it cannot be done. I wrote a letter at that time to Prabhat Sankar Mishra. It is happening like this and what is the redressal. He treated that letter like a public interest petition and in his judgment he commented that the Government of India abdicated their authority to the whims of the Election Commission. It is not done and it can not be done. Why I am mentioning this here is that even if the Government wants to allow export of cotton, the bureaucracy comes in the way and they try their level best to see that it is not done and it is delayed. Therefore, they routed it through Election Commission to delay the process. So these are some of my observations as far as cotton farmers and their problems are concerned. I request you to include these three points *i.e.*, opening a Government Depot for pesticide sale to dealers, allowing cotton under Open General License export and the declining trend in the production of agriculture in Andhra Pradesh in the conclusions of the Seminar.

About the seed aspect, I am told in Punjab the seed companies like the ITC or Mahico or Mahindra the Multi-national companies are non-existent. They could not have a foot hold in the entire Punjab. State. Why it is happening? I am told that the Marketing Federation of Punjab as well as the Agricultural University of Punjab are supplying seed to them and in addition to that they are not that crazy to go to the hybrid. They are going to the State varieties and there is no incentive for the seed manufacturer or seed dealer to adulterate the State varieties. Another aspect that I would like to suggest to the Agricultural University and the agricultural department is to educate the farmers, to produce their own seed. Unless it is done it is a Himalayan task to the Government or Governmental agencies to check seed adulteration. It is not a big job for the farmer to manufacture his own seed, when a company is doing the same by engaging some illiterate agricultural labour to produce hybrid seeds, I hope the enterprising farming community in Andhra Pradesh can rise to the occasion provided the technical know-how being provided by the Agricultural University and the agricultural department. These are some of my observations and I am thankful to Dr.Mohan Rao and the Department of Cooperation and Applied Economics for providing me an opportunity to share my views with all of you.

Dr. Prasada Rao

Seed is a very crucial component and nothing will come without seed. Therefore, the New Seed Policy of encouragement of private seed producers and discouraging the public seed companies, *i.e.*, the State Seed Development Corporation is hurting; particularly in case of cotton, where the entire seed distribution is in the hands of private trade. There is no Government seed and we are seeing the disastrous consequences. In case of rice where varieties are there, there are no hybrids and the State agency is involved, the problem is not that serious. May be I or 2 per cent of seed is substandard or failure of germination. By and large the seed is good. So this Seminar can recommend that the Seed Act should be strengthened in favour of the farmers, and seed production and distribution as far as possible be at the village level. Then even hybrid parents also can be given to the farmers. University can prepare the parents or some state agency can prepare parents of the seed : Hybrid seed they can supply so that seed can be prepared at the village level it self. After all, let it be decentralised below the seed companies. Unless the seed companies are by passed there can not be much relief to the farmers. Therefore, from this Seminar we can say that even hybrid seeds also let the parents be multiplied led by a seed agency under public distribution and the parent seeds are supplied to the villages. The farmers can prepare their own seeds and if he can use, it will be highly useful.

Regarding pesticides IPM has to be adopted. There is no compromise on that. We have to discourage use of pesticides. Unfortunately, Andhra Pradesh has got the dubious distinction of being highest pesticide consuming State in the country and we have to discourage this. For IPM, it requires lot of strengthening of extension department and also the training facilities for the farmers. So from this Seminar we can recommend to strengthen the agency to implement the IPM particularly at the local level, strengthen the training facilities for farmers on IPM at least at every mandal level. That will go a long way. I think we should add these two recommendations.

Coming to credit, it is a major item. Without capital there will not be any capitalistic mode of production. Therefore, the credit cards which the Central Government has announced should be simple enough to take the requirement of not only the cost of production but also to tide over small family needs. If they can be linked to marketing and purchase of inputs it will go a long way in improving the production. So provision of credit cards and linking it to input purchase and then output purchase will give a lot of relief to the small farmers. Another important point is, fertiliser is not a single commodity in the same sense of the economists. They feel that fertiliser is one commodity, one input but what is happening is there are different kinds of fertilisers. There should be balanced fertilisers, Nitrogen, Phosphorous, Potassium and micro nutrients. They are required in small quantities. Pure material is not available. There are no reliable manufacturers. Some reputed companies like Nagarjuna or some other companies can be encouraged to manufacture Zinc Sulphate and other micro nutrients. Such critical components of production which are required in small quantities but which are very crucial at least, the public distribution should be involved. Otherwise there is a lot of adulteration and overpricing that is going on.

Prof. R. M. Mohan Rao

One word Dr. Prasada Rao, you are talking of public agency. A week back, outdated pesticides were caught in Anakapalli Marketing Society and the salesmen were arrested. These days, there seem to be no difference between private and public agencies.

Dr. Prasada Rao

If it is a public agency we can at least ask. In case of private agency if we ask, he may engage goondas to take care of us. In one instance, farmers filed a case in consumer court against seed companies. The private seed companies engaged goondas and forced the farmers to withdraw the case. A public company cannot do such things and accountability will be atleast more In a public agency than a private one.

Dr. Y. Shivaji

The general reality is, in all countries throughout the globe, governments are withdrawing from economic activity. Therefore., even if you want the government should step in distribution, it may not be possible. In that context what I suggest is that let it be freely imported. Some semi-government organisations like, IFCO, may be asked to supplement the micro nutrients.

Dr. Pradada Rao

We are hearing of bio-pesticides. They are not within the reach of farmers, not available to him.

Prof. Ch. A. Krishna Rao

Just I have one comment to make on what Dr. Shivaji has said. He has suggested that we should allow free export of cotton to other countries because the prices for cotton in the international market is about 15 per cent more than the domestic price and the industry is benefited because of not allowing exports of cotton to other countries. What I feel is that if we allow free export of cotton, the price of cotton will increase and the industry people will not keep quite. They will raise the price of cloth and we the consumers will be affected and also our export of cloth will be affected. So these considerations should be borne in mind.

Dr. Shivaji

That is why I mentioned the figures also. Among the total final proceeds of Rs. 960, the share of the raw material supply is as low as Rs.50. So it does not matter much whether it is Rs. 50

or Rs. 100 in the total proceeds of Rs. 960. The textile lobby is capable enough to conserve some money in other areas. These Rs. 50 or 40 does not matter much and second aspect is why should farmers be penalised if they can get a higher price some where else. Why I should not be allowed to reap that benefit when the input which is superior in quality is available outside. On both sides I am affected I am not allowed to reap the benefit by importing the quality input at a cheaper price by protecting the domestic industry and at the same time I am not allowed to reap the benefit of getting a higher price to my output. That way also I am affected. So the true liberalisation means the industry should learn, the rules of the game. Now we are with liberalisation and everybody is pleading for liberalisation in every industry and liberalisation does not mean giving up. Liberalisation means that competition is there. Therefore you should compete with others and see that it goes on. But the protection to the industries the singular failure of the economy for the last 50 years to my knowledge is whether the economists agree with me or not but industry is over protected in this country. That is the singular failure of the economy that we should know. But protection should go.

Prof. R.M. Mohan Rao

Just I will make only one comment. While I agree with most of the suggestions of Dr. Shivaji, I wanted to make one observation on his suggestion that Government should supply seed and pesticides or through some other agencies. These are only palliatives. They will not really hit at the problem. Because what happens is let us take what Shivaji has suggested. Let the Rallies Company keep its products in the Government Depot. Then let us see what will happen. If you are familiar with the literature, that has come out, in Warangal the same tins are used to market the adulterated products. That is the major problem.

18

Concluding Observations

Causes for Suicides

The Seminar has discussed in detail the alternative perspectives on farmer's suicides. There was unanimity on the role of adulterated inputs, poor extension services, lopsided inflow of institutional credit and indebtedness in this regard. But opinions slightly diverged on the liberalization perspective. However, the observation of the NABARD Unit's study that accumulated indebtedness coupled with high degree of mismatch between the expected and actual yields of cotton as the dominant factor contributing to suicides was well received. There was also a broad consensus that the waning out of joint family system on the one hand and declining community concern for others misery on the other have partly been responsible for the diaspora of suicides. Psychological aspects like stress which an individual in unusual situations is confronted with has emerged as another factor contributing to suicides.

Measures to Arrest the Problem in Future

There was a broad consensus on the view that situations of mass suicides of this nature should be treated on par with natural calamities and preventive and curative steps should be taken up. The seminar also felt that the State should not absolve itself of the responsibility in the name of liberalisation and/or globalisation. East Asian nations like Republic of Korea which traversed on the path of market economy never absolved themselves of their responsibility to the farm sector.

The prospects of cotton in the context of liberalisation policies and continued expansion in the traditional areas and extension in the new areas, despite suicides of fellow farmers call for measures to sustain cotton cultivation in a way that the white gold showers prosperity on all the concerned and the Country at large.

Prof. C.H. Hanumantha Rao's observed that the same cotton which has brought misery to many farmers, if handled appropriately can help bring prosperity to the poor and could be a potential tool for eradication of rural poverty was well received. It is against this perception of the role of cotton in the farm economy and to the Nation, the Seminar took stock of the measures to make cotton farming a viable proposition.

For convenience, the recommendation that crystalised from the deliberations are grouped and presented in relation to what is expected of at institutional and individual levels.

(i) Central Government

The role of the Centre is perceived in terms of its regulatory and promotional functions. Regulatory role pertains to seeds and agricultural chemicals, in particular. Cotton, though accounts for 5 per cent of cropped area in the country, consumes more than 55 per cent of pesticide use in the country and this accounts for around 1/3rd of the paid out costs of cotton per hectare. In view of the above, the Centre, should plug in the shortcomings in the Acts dealing with seeds and pesticides. It is also felt that it is reasonable to make it mandatory a Degree in Agriculture—B.Sc. (Agri.) to become a licensed Agro-Pharmacist.

The Seminar noted with concern the quite inadequate allocations to agricultural research *viz.*, 0.5 per cent of the G.D.P. In view of the prospects for making cotton a profitable proposition, the potential and promising developments in the bio-technology needs exploitation to generate pest-free or pest-resistant seeds as is being done in the U.S.A. Exploiting these opportunities calls for greater allocations for scientific research. In this process, the Centre should also explore the possibility of making it mandatory for associations of cotton mills and textiles to allocate at least

1 per cent of their profits for basic research in cotton on a coordinated approach.

The risky nature of cotton farming and the entry of the low riskbearers into this high-risk-agriculture calls for immediate and effective mechanism. Manufacturers of seed and agro-chemicals enjoy assured business, while farmers alone bear the risk of crop loss. These manufacturers should be made to share a part of the risk, by contributing to the cotton insurance pool.

The Seminar felt liberilisation policies have affected agricultural sector in the form of declining share of bank credit without any freedom to export farm products. In this regard, the need to bring cotton exports under OGL was recommended to provide higher prices to farmers.

(ii) State Government

As noted, regulative and effective measures relating to sale of seed and plant protection chemicals should receive top priority, as agri-chemicals are like medicines they will yield the desired results if they are of the right quality and it can be deadly if otherwise. Andhra Pradesh has the distinction of consuming the highest per hectare pesticide consumption in the Country. In this context appointment of agro-pharmacists, on the lines of pharmacists be made mandatory for opening agri-chemical depots. Regulation of agri-chemical traders, given the system lapses and human weaknesses, has its limitations. This calls for exploring alternatives to the existing channels. This can be taken up by inducing credit cooperatives, market committees, marketing societies, etc., to undertake distribution of these inputs. In addition, it should also initiate steps to train farmers to develop their own seeds and also to train a handful of youth at the village level in methods of easy identifications of adulteration of these inputs.

Extension is an area that needs special attention in the context of emerging trends in high-risk commercial cropping and disproportionate risk. Making use of Information Technology, the State should explore reversing the pattem of limited field staff for extension and top heavy administrative structure. Even this exercise may not help meet the total demand. In this context encouraging NGOs with sound traditions into this area and training few farmers in each village help ease the problem.

Marketing is another area that calls for attention by the State Government. Marketing is still unregulated and unorganized inspite of market fee at I per cent advolerum on the produce being collected from the farmers for the last several years. It is not spent for the development of infrastructural facilities like godowns, communication facilities as well as laying all weather roads linking villages to market place. There is also a view that despite market committees being represented by farmers, the exploitative methods of the functionaries of the market yards still continue to work against the farmer and in favour of the trader. Introduction of electronic scales and computerisation of market transactions wherever possible help reduce unfair trade practices. The same could be followed to display the prevailing price at the market yard. Providing free access to market information through electronic media and/or computerised information system, as in the case of Republic of Korea, where free access to recorded ruling prices in any market in the Country, helps selection of right market and timing of sale of product by the farmer. The costs in this regard can be met though market cess collected in the market yard.

It is common knowledge that in its present form the State departments of agriculture are not equipped for providing effective extension to the door steps of the farmers. They are inadequate even to provide technical know-how to high risk crops like tobacco, cotton and chillies. In this context, there is need for shift from over administration and poor extension. The Mandal level Agricultural officer shall have to be entrusted with duties to report pest levels in each village and it should be his primary responsibility. Adequate demonstration farms to impart technical know-how through IPM Schools should receive priority.

Seeds and pesticides should receive prior attention of the State Agricultural Department. They can be relieved of distributional roles like supply of sprayers on subsidy basis which can as well be handed over to non-technical organsiations like market committees or similar appropriate organisations. The

Departments should acquire breeder seeds from agricultural research stations and develop quality seeds through their own seed-farms for distribution among farmers through market committees. They should also help farmers to develop their seeds. In consultation with agricultural scientists of research stations the Department of Agriculture should disseminate information on pests, intensity, timing and method of application of pesticides in addition to exercising constant vigil over spuriousness of pesticides in their areas. Updating their knowledge through constant interaction with agricultural research stations and Agricultural Universities through short-term refresher courses should receive top priority.

(iii) Role of Lead Banks

Annual Credit Plans being prepared by the Lead Banks should take cognizance of the relative share of cropped area under various crops also in fixing targets under the S.A.O. loans. This facilitates flow of credit to crops in proportion to their area. To augment credit flow particularly for commercial crops involving higher working capital the possibility of grouping marginal and small farmers into SHG should be explored. They should also take the assistance of the NGO's working in the area in organising SHG's among marginal and small farmers.

(iv) Agricultural Universities

Agricultural Universities can play an effective role in making available genuine/quality seed through limiting varieties and by undertaking genetic purification of popular varieties grown in different parts of States. Similarly, they could play an active role in limiting the use of chemical pesticides and popualarising bio-pesticides by effectively coordinating their efforts with agricultural departments in transmission of technical know-how and organising demonstration plots. Agricultural Universities can also explore developing seeds which are resistant to the most devastating pests.

(v) Non-Governmental Organisations (NGOs)

NGOs can supplement or function in area where extension

machinery is scarce or not equipped for the task. Creating awareness among farmers on their rights as consumers of farm-inputs. They can also spread information on cost saving and improved technologies besides arranging for testing of soil, seed, pesticides and fertilizers. The NABARD's lessons on the role of NGOs in Khammam in this regard could be of some value. Availability of biological elements is one of major problems for adopting IPM in India. There are also several extension gaps. Education of farmers on the importance of IPM is another area NGOs can explore.

(vi) Vikas Volunteer Vahini Clubs

Credit flow through institutions is at a low pace in the districts under study. VVV Programmes which are organised on the basic premise of development through credit could be of help for greater flow of credit as well as its proper utilisation and repayment.

VVV clubs also could be of some help to the farmers in the percolation of technical know-how and check the distribution of spurious seeds through drawing the attention of the authorities concerned and acting as pressure groups both on those affecting the interests of the farmers on the one hand and in disciplining the farmers attempting to affect the larger interests of credit institutions on the other.

(vii) Farmers

In the final analysis, those affected must try to protect themselves through proper use of the infrastrucure created by governmental and non-governmental agencies. In this context, organising themselves into groups helps their bargaining strength for their just cause. Organisation of compact crop-wise informal groups for credit and marketing comprising kith and kin, relatives and friends numbering between 25-50 in each of the villages helps fruitful interaction with these institutions. Such organisations need not be formal and could be spontaneous centred around a common cause. Secondly, to the extent possible farmers should heed to the advise of the Department of Agriculture in matters relating to use of pesticides. In farm planning and budgeting also they should be cautious in taking risks highly disproportionate to their means.

Prof. R.M. Mohan Rao

The session has come to a close. All the three sessions have been inspiring. I thank the Chainnan, Prof. G.Parhasarathy, Prof. M. Jagadeswara Rao and Dr. Y. Shivaji for their contributions as well as conducting the proceedings in an exemplary manner.

We are grateful to Prof C. H. Hanumantha Rao for setting the tone for discussion as well as for his active participation, in the deliberations of the Seminar. He traced the diaspora of suicides to disproportionate risk taken by the small and marginal farmers and the absence of institutional machanisms to absorb the risks. He insisted that this risk should be shared by other beneficiaries like the input dealers as well as the textile trade the beneficiaries from cotton production. He advocated that liberalisation policies should not come in the way of extending institutional credit to the farmers, particularly the small and marginal ones. He suggested that marginal rise in interest rates or subsidising the banks to compensate for costs involved in providing small amounts to the small and marginal farmers. His observation and suggestion go a long way in reducing the misery of the small and marginal producers of cotton. Finally, Dr. Rao's vision that the same cotton which has brought misery to the farmers can usher in prosperity and help reduction of poverty if properly handled, goes a long way in influencing policies relating to cotton to make it a viable proposition.

During the proceedings various aspects of risk born by the cotton growers have been discussed at length from different disciplinary perspectives. Several measures have been suggested to meet the challenges of risk.

I am grateful to all of you for your active participation and valuable contributions in enriching the proceedings of the Seminar.

List of Participants

1. **Prof. C.H. Hanumantha Rao**
   Honorary Fellow,
   Center for Economic and Social Studies,
   Hyderabad. (A.P.)
2. **Dr. R. Radhakrishna**
   Vice-Chancellor,
   Andhra University,
   Visakhapatnam.
3. **Prof. B. Sarveswara Rao**
   Former Vice-Chancellor,
   Nagarjuna University,
   Visakhapatnam.
4. **Prof. G. Parthasarathy**
   Director, Institute of Development and Planning Studies,
   MVP Colony,
   Visakhapatnam.
5. **Dr. Yalamanchili Shivaji**
   Ex-M.P.,
   Member Central Cotton Advisory Board, Guntur.
6. **Dr. K.J.S. Satya Sai**
   Asst. General Manager, Dept. of Economic Analysis and Research, NABARD, Jeevan Seva Complex (Annex) S.V. Road, Santacruz (W),
   Mumbai-400054.
7. **M.S. Sardesh Pande**
   Asst. General Manager Development (Lead Bank),
   State Bank of India Local Head Office, Bank Street,
   Hyderabad-500095.
8. **Janaki Rama Raju**
   Asst. General Manager,
   State Bank of India,
   Local Head Office,
   Hyderabad.
9. **Dr. A. Prasada Rao**
   Radio Tracer, Agricultural Chemist
   Acharya N.G. Ranga Agricultural University,
   Hyderabad-500030.
10. **Sri G. Sambaiah**
    Farmer Peda Gottipadu,
    Guntur District.
11. **Prof M. Jagadeswara Rao**
    Director Agro-Economic Research Centre and Dean, Academic Affairs, Andhra University, Visakhapatnam.
12. **Prof. B. Prasada Rao**
    Retired Professor, Dept. of Cooperation and Applied Economics, Andhra University, Visakhapatnam

13. **Prof. S. Rami Reddy**
    Retired Professor, N.C.S. Unit, Dept. of Cooperation and Applied Economics, Andhra University, Visakhapatnam.
14. **Prof. V. Bala Mohan Das**
    Principal, College of Arts & Commerce, Andhra University, Visakhapatnam.
15. **Dr. K.S. Chalam**
    Director, Academic Staff College, Andhra University, Visakhapatnam.

Department of Botany

16. **Prof. V. Padmaja**
17. **Dr. A. Arundhati**
18. **Dr K. Uma Devi**
19. **Dr. K. Aruna Lakshmi**

Department of Sociology

20. **Dr. M. Indira Devi**
21. **Mr. K.V. Naidu**
22. **Prof. V.S. Bose**
    Dept. of Psychology, Andhra University, Visakhapatnam.
23. **Prof. K.V. Suryanarayana** Dept. of Statistics, Andhra University, Visakhapatnam.
24. **R.Vijaya Lakshmi**
    Helpage India, Visakhapatnam.

Department of Economics

25. **Prof. P.V. Sharma**
    Director, School of Economics .
26. **Prof. Ch. A. Krishna Rao**
27. **Prof. K. Shiva Prasad**
28. **Prof. G. Ramachandrudu**
29. **Prof. V. Gandhi Babu**
30. **Prof. K. Sreerama Murthy**
31. **Dr. U.N. Murthy**
32. **Dr. R. Sudarsana Rao**
33. **Dr. T. Koteswara Rao**
34. **Dr. P. Tara Kumari**
35. **Ms. M. Sudha**
    Research Scholar

Department of Cooperation and Applied Economics

36. **Prof. R.M. Mohan Rao**
    Chair Professor, N.C.S. Unit
37. **Prof. P. Leela**
38. **Prof. D. Haranatha Babu**
39. **Prof. N.D. Raj**
40. **Prof. D.S. Prasad**
41. **Dr. C.S. Murty**
42. **Dr. K. Venkata Reddy**
43. **Dr. R. Ramakrishna**
44. **Dr. N. Bhaskara Reddy**
45. **Dr. Ch. Appa Rao**
    Research Officer, N.C.S. Unit
46. **Dr. N. Subba Rao**
    Research Officer, N.C.S. Unit

47. **Mr. M. Chalapathi Rao**
J.R.F.N.C.S. Unit
48. **Mr. M. Krishna Mohan**
Research Scholar
49. **Mrs. B. Padma Narayana**
Research Scholar
50. **M. Meles Gebru**
Research Scholar

INDEX